T0073773

Crackling Noise

Crackling Noise

Statistical Physics of Avalanche Phenomena

Stefano Zapperi

Department of Physics and Center for Complexity and Biosystems,
University of Milan, Italy

OXFORD
UNIVERSITY PRESS

OXFORD
UNIVERSITY PRESS

Great Clarendon Street, Oxford, OX2 6DP,
United Kingdom

Oxford University Press is a department of the University of Oxford.
It furthers the University's objective of excellence in research, scholarship,
and education by publishing worldwide. Oxford is a registered trade mark of
Oxford University Press in the UK and in certain other countries

Published in the United States of America by Oxford University Press
198 Madison Avenue, New York, NY 10016, United States of America

British Library Cataloguing in Publication Data
Data available

Library of Congress Control Number: 2022930878

ISBN 978–0–19–285695–1

DOI: 10.1093/oso/9780192856951.001.0001

Printed and bound by
CPI Group (UK) Ltd, Croydon, CR0 4YY

To Caterina

Preface

Materials properties are very often affected by the presence of internal inhomogeneities and structural defects. Hence, the response of materials and the functioning of devices is often associated with noise. In this book, we concentrate on a particular type of noise, known as crackling noise, that is often observed in materials subject to external perturbations that can be of mechanical, thermal, electrical, or magnetic nature. The term crackling noise was introduced by Jim Sethna to describe intermittent series of pulses of broadly distributed amplitude and duration that are characteristic of the sounds we hear in front of a fireplace when wood burns or as we unwrap a piece of candy. It turns out that crackling noise is very common and can be observed in a wide spectrum of conditions and at a broad range of scales. Common examples are the sound we hear when we tear some paper, when we drop sand grains on a table, or when milk pops when entering the pores of our favorite cereals. Well-studied examples of crackling noise in condensed-matter physics are the dynamics of ferromagnetic domains during magnetic hysteresis or the motion of magnetic vortices in superconductors. The electric or magnetic noise emitted by these materials is statistically very similar to the sound emitted by wood burning in a fireplace. On much larger scales, a sequence of earthquakes emits acoustic signals that share many of the features associated with crackling noise in materials.

While representing a nuisance in many practical applications, crackling noise can also tell us something useful about the microscopic processes ruling the materials' behavior. Each crackle in the noise series usually corresponds to a localized impulsive event—an avalanche—occurring inside the material. The broad distribution of pulses in the noise signal reflects the presence of a wide spectrum of avalanches, either localized in a small region, or spreading across the entire sample. A distinct statistical feature of crackling noise, and of the underlying avalanche behavior, is the presence of scaling, observed as power-law-distributed noise pulses, long-range correlations, and scale-free spectra. These are the hallmarks of critical phenomena and phase transitions. Hence, the physics of complex non-equilibrium disordered systems provides the natural theoretical framework to tackle crackling noise.

Theoretical physicists have developed a large set of models, analytic tools, and theoretical ideas that are perfectly tailored to study crackling noise and avalanche phenomena. Relevant examples discussed extensively in this book include early models of self-organized criticality, the theory of elastic manifolds in disordered media, disordered spin models, and percolation. The concept of scale invariance and the renormalization group method provide the guiding principles for our current understanding of avalanche phenomena. Finite-size scaling methods allow for a very precise determination of scaling exponents from numerical simulations, but also of entire scaling functions describing the shapes of crackling noise pulses and the associated avalanche distributions.

Beside its fundamental interest, crackling noise is important for the development of non-destructive testing devices, failure-monitoring tools, and hazard-prediction methods. The viability of such methods relies on appropriate statistical tools acting as indicators of the internal dynamics and microstructural details of the system at hand. The issue is particularly complex because the interplay between long-range interactions, due for instance to internal stresses or electromagnetic fields, and structural disorder can give rise to an amplification of the fluctuations in the material response. In macroscopic samples, noise and fluctuations are typically averaged out: Materials response and device functioning can be described by continuum equations and constitutive laws, as commonly used in engineering calculations. When the system size is reduced toward the nanoscale, however, the traditional methods are bound to fail. Small systems are typically associated with fluctuations that are large as compared to the mean, so that homogenization methods are ineffective. We thus need a theoretical formalism that can treat explicitly fluctuations in the constitutive response of novel nanoscale materials and are able to deal with a sample below the homogenization scale. One could reduce fluctuations by decreasing the degree of disorder in the sample. This, however, may not be effective since even very weak disorder may be amplified by system-spanning interactions, and it is thus necessary to understand the physical mechanism underlying the fluctuating response. In other words, crackling noise can not be avoided or ignored. Our best option is to understand it.

The goal of the present book is to summarize our current understanding of crackling noise, reviewing research performed in the past 30 years, from the early and influential ideas on self-organized criticality in sandpile models, to more modern studies on disordered systems. The book will cover the main theoretical models used to investigate avalanche phenomena; describe the statistical tools needed to analyze crackling noise; and provide a detailed discussion of a set of relevant examples of crackling noise in materials science, including acoustic emission (AE) in fracture, strain bursts in amorphous and crystal plasticity, granular avalanches, magnetic noise in ferromagnets and superconductors, and fluid flow in porous media. Finally, one chapter is devoted to a discussion of avalanches in biological systems.

The book is conceived as an introductory monograph to the field of the physics of crackling noise. Its targeted readership is composed by advanced graduate or master's students in physics, materials science, and engineering, especially those interested in statistical methods for materials. The book should also appeal to established researchers in those fields and to others who wish to learn basic notions on avalanche phenomena. To make the book accessible to students, an effort has been made to provide introductory sections to recall basic concepts and ideas (e.g. percolation, elasticity, critical phenomena) that can be useful to understand the more complex topics treated throughout the book.

Acknowledgements

The genesis of this book has a long history spanning more than two decades. I was involved in crackling noise and avalanche phenomena from the very beginning of my scientific career, as an undergraduate student at the University of Rome 'La Sapienza" in the group of Luciano Pietronero, who proposed that I work on a real space renormalization group for sandpile models. Later I continued working on avalanches in disordered media as a Ph. D. student in the group of Gene Stanley at Boston University. A first proposal for a book on "Self-organized criticality and avalanche phenomena" was conceived with Alessandro Vespignani almost 20 years ago. The idea to write a book on this subject stemmed from our long-lasting collaboration on the fundamental properties of self-organized criticality, but despite the enthusiasm from the publisher and some preliminary work, the project never really took off and was finally abandoned. I am nevertheless indebted to Alessandro for endless discussions and intense collaborations on self-organized criticality, fracture, and dislocations that shaped my understanding of this field.

During my career, I was fortunate to collaborate with several outstanding colleagues who contributed greatly to my understanding of the topics discussed throughout the book. The list is extremely long and I am hoping not to forget anyone. I first wish to thank Jim Sethna, who should be credited for introducing in the scientific literature the term "crackling noise" used as the title for the present book. I also thank Jim for the many enlightening discussions and enjoyable collaborations on many of the topics included in this book, such as plasticity, fracture, and magnetism. The work on self-organized criticality and sandpile models that is discussed in this book was done mostly in collaboration with Mikko Alava, Ronald Dickmann, Lasse Laurson, Kent B. Lauritsen, Miguel A. Muñoz, and Alessandro Verspignani. My views on the Barkhausen noise result from my decades-long collaboration with Gianfranco Durin, who performed the most accurate experimental measurements for this type of crackling noise. I also acknowledge discussions and collaborations on the same topic with Giorgio Bertotti, Felipe Bohn, Claudio Casellano, Pierre Cizeau, Francesca Colaiori, Andrea Gabrielli, Alessandro Magni, Stefanos Papanikolaou, and Ruben Sommer. I approached the field of crystal plasticity 20 years ago thanks to Jerome Weiss and Jean-Robert Grasso, who performed insightful experiments on the creep deformation of ice single crystals. I also had the pleasure to collaborate on the same project with Carmen Miguel, with whom I continued to explore the properties of dislocations in regular crystals and in vortex lattices. She read preliminary versions of a few of the chapters of the present book and always encouraged me to complete it. At around the same time, I had the chance to meet with Michael Zaiser, who became my reference figure in the field of dislocation theory. Our collaboration on the statistical mechanics of crystal and amorphous plasticity is still ongoing after more than 20 years. In the field of dislocation dynamics, I also acknowledge collaborations with Dennis Dimiduk, István Groma, Péter Ispánovity, Markus Ovaska, Stefanos Papanikolaou, and Daniel Weygand. My grasp amorphous plasticity was influenced by more recent discussions and collaborations with Silvia Bonfanti, Ezequiel Ferrero, Roberto Guerra, Itamar Procaccia and Stefan Sandfeld. I also wish to thank José Soares Andrade and André Moreira for our collaboration on vortex matter.

My work on fracture was mostly done together with Mikko Alava, with whom I also collaborated on other problems discussed in this book, such as sandpile models, the random field Ising model, and dislocation dynamics. I also thank Mikko for his warm hospitality for many months in Finland. In the area of fracture, I should also credit discussions and collaborations with Guido Caldarelli, Hans Hermann, Ferenc Kun, Ian Main, Phani Nukala, Stephane Roux, Stephane Santucci, and Ashivni Shekhawat. My research activity on granular media benefited from collaborations with Andrea Baldassarri, Fergal Dalton, Hans Herrmann, François Lacombe, and Alberto Petri. I wish to thank my students and postdoctoral fellows that worked with me on several problems discussed in this book. In particular, I would like to mention Andrea Benassi, Zsolt Bertalan, Zoe Budrikis, Benedetta Cerruti, Oleksandr Chepizhko, Giulio Costantini, Francesc Font-Clos, Fabio Leoni, Manuela Minozzi, Paolo Moretti, and Alessandro Taloni. I acknowledge Dennis Dimiduk, Gianfranco Durin, Lasse Laurson, and Michael Uchic for kindly providing images or data for some of the figures in this book. The book also benefited from many conversations with several other colleagues that I would like to thank here: Daniel Bonamy, Elisabeth Bouchaud, Jean Philippe Bouchaud, Karin Dahmen, Karen Daniels, Thierry Giamarchi, Alex Hansen, Pierre Le Doussal, Craig Maloney, Alberto Rosso, Lev Truskinovski, Damien Vandembroucq, Eduard Vives, Kay Wiese, and Matthieu Wyart.

Parts of this book were written in Germany, thanks to the generous support of the Alexander von Humboldt foundation that supported my sabbatical stay at FAU Erlangen-Nuremberg and LMU Munich through the Humboldt Research Award. I thank Michael Zaiser and Erwin Frey for their hospitality during this period of time. I would also like to thank Sonke Adlung at Oxford University Press for relentlessly pushing me to write a book on avalanche phenomena. He persisted for almost two decades in his belief that a book like this had to be written.

Last but not least, I wish to warmly thank Caterina La Porta, my beloved partner in life and essential coworker in science. The little I know about biology, I owe it to her. She should be credited for the main ideas discussed in the last chapter of the book on avalanches in biological systems, although any mistakes or inaccuracies are fully my responsibility. Without her constant encouragement over several years, this book would never have been completed.

Contents

1

Scaling Features of Crackling Noise

In this chapter, we discuss the statistical properties of crackling noise using a simple sandpile model as an illustration. We recall some basic concepts related to the theory of phase transitions and critical phenomena and then show how they can be applied to the description of avalanche phenomena. In this context, we discuss critical exponents, scaling relations, and universal scaling functions using the Ising model as an example. We discuss the spectral properties of crackling noise and show how they can be related to the scaling of the avalanches. Finally, we discuss the effects of driving rate and background noise on the avalanche statistics.

1.1 Crackling noise

A wide variety of very common physical phenomena produce a crackling type of noise, characterized by intermittent large bursts followed by quieter periods (Sethna *et al.*, 2001). Think for instance of the sound emitted by wood in a fire, the noise we hear when we unwrap a candy bar or tear a piece of paper. Similar crackling sounds can also be heard when we listen carefully to the bubbles bursting in a glass of champagne or rice cereals popping in a bowl when milk is added. On a much larger scale, if we were to listen to the sound made by earthquakes we would hear again a similar crackling activity.

While crackling noise is very common, not all noises are of this type. For instance, white noise is an unstructured, often disturbing, hissing sound very different from crackling noise. The name white noise comes from an analogy with white light, a uniform mixture of light with different wavelengths. Similarly, white noise is a uniform mixture of sounds with different frequencies. If we plot the signal intensity $V(t)$ of white noise, we observe a very random pattern like the one reported in the top panel of Fig. 1.1. White noise has a uniform spectral density: Its power spectrum $F(\omega) = |\hat{V}(\omega)|$, defined as the squared amplitude of the Fourier transform of the signal intensity, is a constant.

When the noise spectrum is not constant, we talk about "colored noise." An example of colored noise is the Brownian noise reported in the middle panel of Fig. 1.1 which represents the cumulative integral of a white noise signal or the position of a one-dimensional random walk. The power spectrum of Brownian noise decays as the inverse square of the frequency $F(\omega) \sim 1/\omega^2$. Crackling noise is a peculiar form of colored noise and displays a power spectrum decaying as $F(\omega) \sim 1/\omega^a$, where typically the exponent a lies in the range $[1, 2]$. While the spectral properties of crackling noise may look similar or even identical to other types of colored noises, its time series

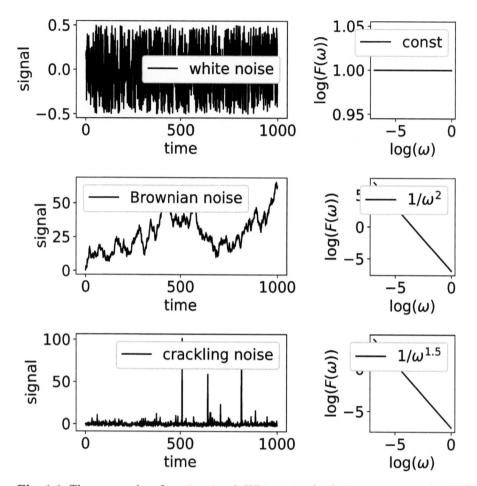

Fig. 1.1 Three examples of a noisy signal. White noise (top), Brownian noise (middle) and crackling noise (bottom). The graphs on the right display the typical power spectrum of the signals. White noise corresponds to a constant spectrum, while Brownian noise and crackling noise have power law decaying spectra.

is drastically different from that of Brownian noise. This is illustrated in the bottom panel of Fig. 1.1 where we can appreciate that crackling noise is defined by its inter-mittent structure, with large peaks followed by smaller ones. From this we see that crackling noise is defined by the presence of discrete pulses or 'crackles."

An important point is that crackling noise pulses originate from underlying physical processes that we call *avalanches*. As the name suggests, an avalanche is a sudden collective event, causally connected in time and space. A very familiar example is indeed a snow avalanche: Due to slowly varying external perturbations, such as an increase in temperature, or to the onset of mechanical triggering events, a block of snow can suddenly detach from a slope and fall downwards. In this example, the duration of the avalanche is much shorter than the time needed by the external perturbations to trigger it. Such *time-scale separation* is essential in defining an avalanche event, since otherwise we cannot distinguish one avalanche from the next.

Understanding the physical mechanisms underlying avalanche propagation is not only interesting for geologists, but this behavior is common to many physical phenomena. Indeed, in this book we will discuss only tangentially geophysical phenomena, such as snow avalanches, landslides, and earthquakes, and focus instead on examples taken from condensed matter physics. These include among others fracture and crack propagation, dislocation dynamics in crystal plasticity, the Barkhausen effect in ferromagnetic materials, the dynamics of flux lines in type II superconductors, fluid flow through porous media, and granular friction. The common feature of all these systems is a slow external driving causing an intermittent, widely distributed avalanche response, with few large events and many more small events. The study of avalanches in statistical physics was pioneered by the work of Bak *et al.* (1987) who proposed that avalanches with broad power law distributions would reflect the presence of a self-organized critical point. To illustrate their ideas, they introduced a simple discrete model representing a sandpile. We will follow the same track to illustrate the statistics of avalanches in crackling noise, but first we take a brief detour to recall some basic concepts related to critical phenomena. Readers who are familiar with the theory of phase transition and critical phenomena can safely skip the following section.

1.2 Scaling and critical phenomena

1.2.1 Phase transitions

According to the self-organized criticality scenario, driven systems should tune themselves precisely at the boundary between two dynamical phases, a static and a moving phase. Avalanches can then be seen as the vivid representation of the transition between motion and stasis. Close to critical points in equilibrium phase transitions, thermodynamic variables obey power law scaling relations and the systems develop long-range correlations. Consider for instance a paramagnetic-ferromagnetic phase transition, where below the Curie temperature T_c the material develops a spontaneous magnetization M, which for simplicity we consider as a scalar quantity, as for uniaxial ferromagnets.

In the language of phase transitions, M is known as the *order parameter* since it describes the ordering of the system, while the temperature is a *control parameter*. As we increase the temperature towards the Curie temperature, the magnetization goes to zero continuously as

$$M \sim (T_c - T)^\beta, \tag{1.1}$$

and remains zero in the paramagnetic phase. Approaching the critical point from the paramagnetic phase, the local magnetization $M(\mathbf{r})$ becomes more and more correlated in space as can be quantified by studying its correlation function which typically at the critical point scales as

$$\langle M(\mathbf{r}')M(\mathbf{r})\rangle \sim \frac{1}{|r'-r|^{d-2+\eta}}. \tag{1.2}$$

where d represents the dimensionality of the system. Away from the critical point the power law decay in Eq. 1.2 is cut off at a correlation length ξ which diverges approaching the critical point as

$$\xi \sim (T_c - T)^{-\nu}. \tag{1.3}$$

Power law relations are also found when we consider the effect of an external magnetic field H: At $T = T_c$, the magnetization increasing with the applied field as

$$M \sim H^{1/\delta}. \tag{1.4}$$

Approaching the critical point, the magnetic susceptibility $\chi \equiv dM/dH$ also diverges as a power law

$$\chi \sim (T_c - T)^{-\gamma}. \tag{1.5}$$

1.2.2   The Ising model

An illustration of the correlations expected close to a critical point can be obtained by studying the Ising model, probably the most studied model of ferromagnetism and phase transitions in general. In the Ising model, a set of spins $s_i = \pm 1$ reside on the site of a d-dimensional lattice. The energy associated with a configuration of spins $\{s_i+\}$ is given by

$$E = -J \sum_{\langle i,j \rangle} s_i s_j - H \sum_i s_i H, \tag{1.6}$$

where $J > 1$ is the ferromagnetic coupling, favoring parallel spins, and H is the external magnetic field (in suitable units). The first sum in Eq. 1.6 is restricted to nearest-neighbor pairs. The Ising model can be solved exactly in $d = 2$ (for $H = 0$) and displays a second-order phase transition at a critical temperature T_c. It is also possible to perform Monte Carlo numerical simulations to inspect the typical equilibrium configurations of the model as the temperature is varied. Examples are reported in Fig. 1.2 for different temperature values, below and above T_c. Notice the change in spin cluster size as the temperature is varied. At $T = T_c$, correlations are long-ranged and we see clusters of all sizes.

1.2.3   Universality

In writing all these power law relations, we have defined a set of *critical exponents* (i.e. β, η, ν, δ, γ) which play a very important role in describing phase transition. Notice that these exponents are not all independent. It is possible to derive scaling relations linking them. We will later derive some scaling relations between exponents describing avalanche statistics.

According to the theory of critical phenomena, critical exponents are *universal*, meaning that they do not depend on the value of the microscopic parameters of the systems. Thus two ferromagnetic materials with different atomic composition and structure may display the same set of critical exponents close to the transition. The two materials are said to belong to the same *universality class*. The particular universality class of a phase transition only depends on some broad features, such as the

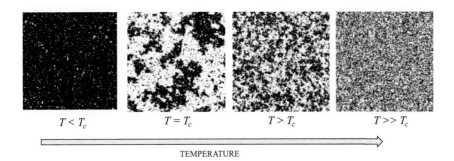

$T < T_c$ $T = T_c$ $T > T_c$ $T >> T_c$

TEMPERATURE

Fig. 1.2 Monte Carlo simulations of the Ising model at different temperatures. Source: Ising model configurations were obtained with the code of Matt Bierbaum (`https://mattbierbaum.github.io/ising.js/`).

symmetry of the order parameter, and the dimensionality. Thus, the phase transition in an isotropic magnet, where the magnetization is a three-dimensional vector, is described by a different universality class than that of the uniaxial magnet described above, where the magnetization is oriented along a well-defined axis and is therefore effectively one-dimensional. Universality is a very powerful concept since it allows us to describe quantitatively very different phenomena without having to deal with the fine details of the microscopic interactions.

Finally, a note about the role of dimensionality in universality. As we increase the dimensionality d of the system, critical exponents change their value, but only up to the upper critical dimension d_c, which in the case of the Ising model is equal to $d_c = 4$. For $d > d_c$, the critical exponents take constant values that can be obtained by mean-field theory, a theory where each element of the system interacts with the mean of the system. To illustrate this point we consider the example of the Ising model. In a mean-field description, the energy in Eq. 1.6 is replaced by $E_{\mathrm{MF}} = -\sum_i (Jm + H)s_i$, where the average magnetization $m = \langle s_i \rangle$ can be obtained self-consistently. Throughout the book, we will provide examples of mean-field theories relevant to describe avalanche phenomena.

1.2.4 Universal scaling functions

In discussing universality, we focused on the value of the critical exponents, which does not depend on irrelevant microscopic details. Universality, however, is an even more powerful concept since it also applies to entire functional shapes. To illustrate this point consider a system with order parameter Φ and reduced control parameters t and h. In the case of the Ising model, Φ is the magnetization M, $\hat{t} = (T_c - T)/T_c$ is the reduced temperature, and h is proportional to the magnetic field . The critical point corresponds to $\hat{t} = 0$ and $h = 0$ where also $\Phi = 0$. The behavior of the order parameter in the vicinity of the critical point can be expressed in terms of a scaling function $f(x)$:

$$\Phi(\hat{t}, h) = t^\beta f(h/\hat{t}^{\delta\beta}), \tag{1.7}$$

where the scaling function $f(x)$ is universal. To satisfy known scaling laws for the order parameter, the function should be constant for $x = 0$, $f(0) = C$, and scale as $f(x) \sim x^{1/\delta}$ for $x \gg 1$. Scaling functions can also be written for the two-point correlation of the order parameter

$$\langle M(\mathbf{r}')M(\mathbf{r})\rangle = C(x = |\mathbf{r}' - \mathbf{r}|, \hat{t}) = x^{-(d-2+\eta)}g(x\hat{t}^{-\nu}), \tag{1.8}$$

where $g(x)$ decays to zero faster than a power law for $x \gg 1$.

A particularly interesting form of scaling functions are those arising in a system confined in a box of finite size L^d. In that case, the correlation length ξ cannot diverge at the critical point but can at most reach the system size L. All the scaling functions should be modified to take into account finite-size effects. For example, the order parameter will scale as

$$\Phi(\hat{t}, h, L) = t^\beta F(h/\hat{t}^{\delta\beta}, Lt^\nu). \tag{1.9}$$

Finite-size scaling functions are extremely useful when using numerical simulations to study critical phenomena. As will be illustrated throughout this book, one typically simulates a model for different values of L and then extrapolates the critical behavior from finite-size scaling assumptions.

1.3   Avalanche statistics

Considered here is a simple lattice model due to Manna (1991), which represents a variant of that introduced by Bak *et al.* (1987) to define self-organized criticality. A more detailed discussion of sandplile models is deferred to the next chapter of this book. In the simple Manna model , individual grains are slowly and randomly added to the sites of a two dimensional lattice. When a single site holds more than one grain, it becomes unstable and *topples*, transferring two grains to two randomly selected nearest-neighbors sites (see Fig. 1.3). A toppling event can then induce nearest-neighbor sites to topple in turn, generating an avalanche. Grains that fall outside of the lattice boundaries are considered to be lost. To enforce time-scale separation, new grains are dropped on the lattice at a very slow rate. In the limit of infinitesimal driving rate , we add grains only when the previous avalanche is over. Due to the balance between the external flux of grains and the dissipation at the boundary, the system will reach a steady state with a constant average density of grains present in the lattice. Avalanches then represent the local fluctuations over this steady-state density of grains.

In Fig. 1.4, we illustrate the spatio-temporal shape of a typical large avalanche in the simple sandpile model. In the left panel, we display the total number of toppling events $V(t)$ occurring at each iteration t during an avalanche. In the right panel, we display the spatial extension of the avalanche, wit the gray-scale indicating the number of toppling events that have taken place in each lattice site. The figure displays on one hand a typical crackling noise pulse, and on the other hand the underlying spatial avalanche associated with the noise pulse.

To characterize the statistical properties of the avalanche, we can define its size s as the total number of toppling events which is equivalent to the area under the pulse shown in the left panel of Fig. 1.4. If we denote by T the duration of the avalanche

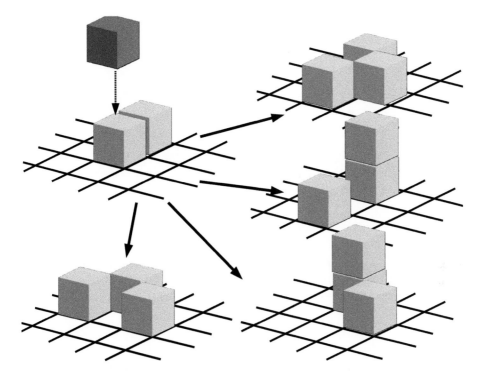

Fig. 1.3 A simple sandpile model introduced by Manna (1991). Grains are added on a lattice. When a site contains more than two grains it becomes unstable and the two grains are transferred randomly to two neighboring sites. The figures shows four possible toppling events. Two of those will lead to further toppling.

pulse, the avalanche size can also be written as $s = \int_0^T dt V(t)$. Furthermore, we can also measure the avalanche area A, as the number of sites that have toppled at least once during the avalanche or the total area shown in the right panel of Fig. 1.4. Clearly, the area of an avalanche is always smaller or equal to its size, since a site can topple more than once.

Avalanches in the sandpile model come in widely different sizes, as illustrated in Fig. 1.5a, reporting a typical time sequence of avalanche events. The intermittency and broad intensity variations of the avalanche series are the hallmarks of crackling noise: Large avalanches are sometimes followed by small events leading to a complex signal. To characterize the avalanche signal statistically, one can compute the distribution of the size of each pulse. For the signal reported in Fig. 1.5(a), the distribution follows a power law up to a characteristic size s_0, which depends only on the lattice size L,

$$P(s) = s^{-\tau} f(s/s_0), \tag{1.10}$$

where $f(x)$ is a cutoff function, decaying exponentially fast, and $\tau \simeq 1.27$ is a scaling exponent. Instead of measuring the avalanche size s, we can also measure the areas A associated with each avalanche. Avalanche areas are also found to be power law

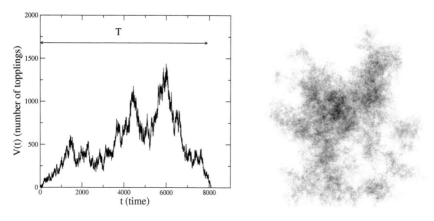

Fig. 1.4 Example of an avalanche in the Manna sandpile model. The number of topplings V as a function of time for the avalanche (left). The spatial extension of the avalanche (right). The grayscale represents the number of toppling for each site (right).

distributed

$$P(A) = A^{-\tau_A} h(A/A_0), \qquad (1.11)$$

where $h(x)$ is another cutoff function, depending on the avalanche characteristic area A_0 and $\tau_A \simeq 1.2$ is a new scaling exponent. Similarly, we can consider the avalanche radius l, which is distributed as

$$P(l) = l^{-\lambda} F(l/\xi), \qquad (1.12)$$

where $F(x)$ is the cutoff function, ξ is the avalanche characteristic length and $\lambda \simeq 1.1$ is the scaling exponent. Finally, the avalanche duration distribution is again described by a truncated power law

$$P(T) = T^{-\alpha} g(T/T_0), \qquad (1.13)$$

where $g(T)$ is a cutoff function, T_0 is the cutoff value and $\alpha \simeq 1.5$. A more quantitative account of the behavior of sandpile models will be considered in Chapter 2; here we use the model just as a means to establish the notation used throughout the book to define avalanche statistics and to derive some general scaling laws.

1.4   Avalanche scaling relations

As discussed in the previous section, the magnitude of an avalanche can be defined in different ways, considering its size s, area A, duration T, or length l. All these quantities are distributed as power law distribution over a wide range of scales, up to a cutoff value. As shown in Fig. 1.5, the cutoff of the avalanche size distribution in the sandpile model increases with the lattice size L, which we can thus define as the characteristic length-scale of the avalanches. All the distributions we have discussed display a cutoff scaling with the lattice size L: the cutoff of the avalanche size scales

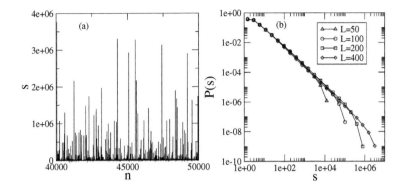

Fig. 1.5 (a) A typical avalanche size sequence for the Manna sandpile model (b) The distribution of avalanche sizes is a power law with a cutoff s_0, that increases with the system size L.

as $s_0 \sim L^D$, the one of the area as $A_0 \sim L^{d_f}$, where d_f is the fractal dimension of the avalanche, and the characteristic avalanche duration scales as $T_0 \sim L^\Delta$, defining another exponent Δ. Such a scale-free behavior, limited only by the system size, was originally considered to be the defining feature of self-organized criticality: a critical phenomenon, without any apparent parameter tuning. We will see in the next chapter how this original idea has evolved.

The fact that the avalanche distributions and their cutoffs are described by power laws allows us to derive scaling relations between the different exponents. We can describe each avalanche by a set of variables $\{s, A, T, l\}$ and then define the joint probability distribution $p_4(s, A, T, l)$, as the probability to have an avalanche of size s, area A, duration T, and length l. The single-variable probability distributions discussed in the previous section can then be obtained by a simple integration (also known as *marginalization* in probability theory). For instance, the size distribution is obtained as

$$P(s) = \int dA dl dT p_4(s, A, T, l). \tag{1.14}$$

If we wish to find a relation between two variables, such as size and duration, we can define the two-point joint probability distribution

$$p_2(s, T) = \int dA dl p_4(s, A, T, l) \tag{1.15}$$

and then compute the conditional expectation value of the size s for a constant duration T

$$\langle s(T) \rangle \equiv \int ds s p_2(s, T). \tag{1.16}$$

In the sandpile model, the average size is found to scale with the duration as

$$\langle s(T) \rangle \sim T^{\gamma_{st}}, \tag{1.17}$$

with $\gamma_{st} \simeq 1.77$. Similar scaling laws are found for all the other avalanche variables: For each pair of variables (x, y), we can define a conditional expectation value $\langle x(y) \rangle$, scaling with as $y^{\gamma_{xy}}$.

To summarize, we have defined a set of exponents to characterize the avalanche statistics:

(i) the exponents describing the power law decay the avalanche distributions τ, τ_A, α, and λ, for size, area, duration, and radius, respectively;

(ii) the exponents describing the scaling of the cutoff as a function of the lattice size L, D, d_f, Δ for size, area, and duration, respectively. The cutoff of the avalanche radius distribution scales linear with the lattice size, so that the exponent is trivially equal to one;

(iii) a set of exponents γ_{xy} defining the scaling of the conditional expectation value of the variable x when y is kept constant.

Notice that all these exponents are interdependent and we can derive relation between them. To this end, we should make the assumption that the two-point distribution is narrow and peaked. Under this asssumption, we can replace the variables by their expected values, imposing that $x \sim y^{\gamma_{xy}}$. By inverting this relation, we can readily derive a scaling relation between different exponents:

$$\gamma_{xy} = \frac{1}{\gamma_{yx}}. \tag{1.18}$$

Expressing x in terms of z and then z in terms of y, we obtain $x \sim z^{\gamma_{xz}} \sim (y^{\gamma_{zy}})^{\gamma_{xz}}$ or

$$\gamma_{xy} = \gamma_{xz}\gamma_{zy}. \tag{1.19}$$

This relations between variables holds also for the cutoff of the distributions, implying a set of scaling relations for the relative exponents

$$D = \gamma_{st}\Delta = \gamma_{at}d_f. \tag{1.20}$$

Finally, we can derive scaling relations for the exponents describing the decay of the avalanche distributions. To this end, we notice that conservation of probability implies that $P_X(x)dx = P_Y(y)dy$, where we have used the notation $P_X(x)$ for the probability that the variable X takes the value x. Changing variables according to $y(x) \propto x^{\gamma_{yx}}$, we obtain $P_X(x) = P_Y(y(x))dy/dx \propto P_Y(x^{\gamma_{yx}})x^{\gamma_{yx}-1}$. If $P_X(x) \sim x^{-\tau_x}$, we can derive a set of scaling relations $\tau_x = 1 + \gamma_{yx}(\tau_y - 1)$. Reverting to the original notation of the avalanche exponents, we can write

$$\tau = 1 + \gamma_{ts}(\alpha - 1) = 1 + \gamma_{as}(\tau_A - 1) = 1 + \gamma_{rs}(\lambda - 1), \tag{1.21}$$

and other similar relations. Using Eq. 1.20 and eliminating the γ_{xy} exponents, we can express the relation only in terms of the cutoff exponents:

$$\tau = 1 + \frac{\Delta}{D}(\alpha - 1) = 1 + \frac{d_f}{D}(\tau_A - 1) = 1 + \frac{\lambda - 1}{D}. \tag{1.22}$$

All these scaling relations are satisfied by the model, as can be checked by inserting the numerical values of the exponents.

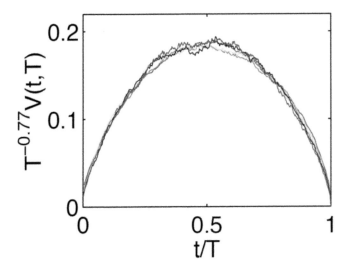

Fig. 1.6 The average pulse shape for the Manna sandpile model. Source: Laurson *et al.* (2005). Reprinted with permission from IOP Publishing.

As discussed in Section 1.2.4, universality is not only restricted to the values of the critical exponents but also to the shape of the scaling functions. In the context of avalanche phenomena and crackling noise, universal scaling functions describe for instance the behavior of the cutoff functions of the avalanche distributions (e.g. in $f(x)$ Eq. 1.10), or the joint distribution of avalanche size and duration. Another scaling function that has attracted some attention is the average pulse shape of avalanches of a given duration (Sethna *et al.*, 2001). For an avalanche of duration T, we have defined $V(t, T)$ as the number of toppling events at time t (see Fig. 1.4). The average pulse shape is defined as the average of $V(t, T)$ over all the avalanches of size T. Since the number of toppling events V scales as S/T, we expect that at the critical point the average pulse shape follows

$$\langle V(t, T) \rangle = T^{\gamma_{st} - 1} \mathcal{V}(t/T), \tag{1.23}$$

where $\mathcal{V}(x)$ is a universal scaling function (Sethna *et al.*, 2001) which is plotted in Fig. 1.6 for the Manna sandpile model (Laurson *et al.*, 2005).

1.5   Spectral properties of crackling noise

Understanding the origin of the the widespread observation of "$1/f$ noise" in natural phenomena was one of the original motivation behind the idea of self-organized criticality (Bak *et al.*, 1987). Early scaling arguments were proposed to derive the shape of the power spectrum of sandpile models (Jensen *et al.*, 1989; Kertész and Kiss, 1990), but a convincing general solution for the power spectrum of crackling noise was only reached later by Kuntz and Sethna. The solution was then shown to be also applicable to sandpile models (Laurson *et al.*, 2005).

As discussed previously, the power spectrum is defined as the square amplitude of the Fourier transform of a signal $V(t)$ or

$$F(\omega) \equiv |\int_{-\infty}^{\infty} dt V(t) e^{i\omega t}|^2. \tag{1.24}$$

It is straightforward to show that for a stationary signal $V(t)$, the power spectrum is also equal to the Fourier transform of the correlation function

$$F(\omega) = \int dt \langle V(t+t_0) V(t_0) \rangle e^{i\omega t}, \tag{1.25}$$

where the average is taken over different values of the starting time t_0.

Since crackling noise can be decomposed into individual avalanche pulses, it is possible to relate the statistics of the avalanches with the shape of the power spectrum. Consider the noise time series $V(t)$, which in the sandpile model corresponds to the number of local relaxation events taking place in the avalanche, defined here as a connected sequence of non-zero values of $V(t)$. For an avalanche of duration T, the size is defined as $s(T) = \int_0^T dt V(t)$. If we assume that the average size scales as in Eq. 1.17, $\langle s(T) \rangle \sim T^{\gamma_{st}}$ and that the dynamics is self-similar, the form of the power spectrum can be obtained by averaging the energy spectrum $E(\omega|s)$ of avalanches of size s. The energy spectrum scales as

$$E(\omega|s) = s^2 g_E(\omega^{\gamma_{st}} s), \tag{1.26}$$

where $g_E(x)$ is a scaling function (Kertész and Kiss, 1990; Kuntz and Sethna, 2000). The power spectrum can then be written as

$$F(\omega) = \int P(s) E(\omega|s) ds, \tag{1.27}$$

where $P(s) \sim s^{-\tau}$ is the probability distribution of avalanche sizes and the integral is bound by the upper cutoff s_0, so that

$$F(\omega) = \omega^{-\gamma_{st}(3-\tau)} \int^{s_0 \omega^{\gamma_{st}}} dx x^{2-\tau} f_{energy}(x). \tag{1.28}$$

If the integral in Eq. (1.28) is convergent, we obtain $a = \gamma_{st}(3 - \tau)$, a relation that was first derived in 1972 (Lieneweg and Grosse-Nobis, 1972) in the context of the Barkhausen noise, more than a decade before sandpile models were introduced.

In most cases, including sandpile models, however, the integral in Eq. (1.28) does *not* converge and the final result crucially depends on the asymptotic behavior of $g_E(x)$. In their early study of the sandpile model, Kertesz and Kiss assumed $g_E(x) \propto 1/(1 + x^{2/\gamma_{st}})$, obtaining $\alpha = 2$ (Kertész and Kiss, 1990). Jensen *et al.* proposed to approximate the avalanche shape with a box function, which implies $g_E(x) \propto (1 - \cos(x^{1/\gamma_{st}}))/x^{2/\gamma_{st}}$, yielding again $\alpha = 2$ (Jensen *et al.*, 1989). Neither of these assumptions is, however, correct for the sandpile model, as pointed out in (Laurson *et al.*, 2005).

Kuntz and Sethna found the correct scaling by noticing that in most cases the avalanche dynamics is a local spreading process. In that case, the released energy must be an extensive function of the size s, or $E(\omega|s) \sim s$. From Eq. 1.26 it thus follows that $g_E(x) \sim A/x$. This implies that when $\tau < 2$, the integral in Eq. (1.28) is dominated by the frequency dependent upper cutoff, yielding $a = \gamma_{st}$. Notice that the relation $\tau < 2$ is satisfied by many examples of crackling noise, including sandpile models, as illustrated in Fig. 1.7.

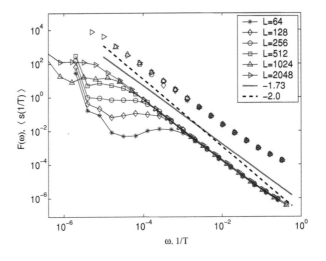

Fig. 1.7 The power spectrum of the two dimensional Manna model for different lattice sizes L. We also plot the conditional expectation value of the avalanche size as a function of $1/T$. The two curves have the same slope in a log-log plot. Source: Laurson *et al.* (2005).

1.6 Driving rate

In the sandpile model, time-scale separation is strictly enforced since no grains are introduced in the system as long as an avalanche is running. This is of course an idealization since in many experimental conditions where crackling noise is recorded time-scale separation is not perfect. It is possible to generalize the sandpile model to take this into account by introducing a non-vanishing driving rate h, defined as the probability for each lattice site to receive a grain at each time step. In the limit $h \to 0$, we recover the results of the sandpile model in the time-scale separation regime. A non-vanishing driving rate implies that different avalanches can merge, perturbing their statistics. If the driving rate is large enough, we would not be able even to distinguish the avalanches, their being replaced by a continuous activity signal. This is illustrated in Fig. 1.8, where we show the crackling noise signal obtained in the sandpile model for different values of the driving rate h.

The effect of the driving rate h on the avalanche distributions was analyzed in general terms in (White and Dahmen, 2003). The results are thought to be generally applicable, as long as the following basic assumptions are satisfied by the system under study:

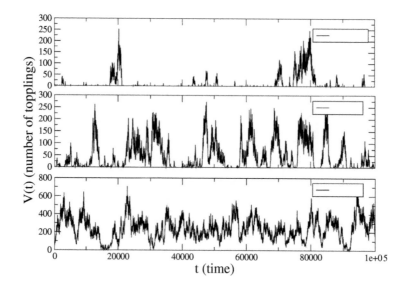

Fig. 1.8 The avalanches in the Manna sandpile model for three different values of the driving rate h. For the smallest rate, avalanches are well separated and approach the slow driving limit. For the intermediate rate, avalanches merge together and their statistics is perturbed. Finally for the fastest rate, it is not possible to identify the avalanches and we are left with a continous signal. Source: Image courtesy of Lasse Laurson.

1. The system is close to a critical point, so that in the adiabatic limit $(h \to 0)$ the size and duration distributions are a power law.
2. The avalanche signal is stationary under the time window considered.
3. In the adiabatic limit, the average number of avalanche nucleation events per unit field increase is a smooth function of the driving field.
4. For low h the avalanche dynamics is independent of h.
5. The field increases between avalanche nucleation events is independent of the avalanche sizes.
6. The avalanche sizes are uncorrelated in time.
7. Avalanches are nucleated randomly in space.

Under those assumptions , it was shown that the effect of the driving rate depends on the value of the exponent α of the duration distribution, in the limit $h \to 0$ (White and Dahmen, 2003). For $\alpha > 2$, the distributions are unaffected by the driving rate , while for $\alpha < 2$ a peak appears in the distribution for large sizes and durations. The most interesting case is $\alpha = 2$, for which we can expect a linear dependence of the exponents on the driving rate h. This general analysis provides an explanation of the results of experiments and numerical simulations, where driving-dependent exponents have sometimes been observed (Durin and Zapperi, 2006). In general, when the driving rate is non-vanishing we should be aware of its effect on the avalanche statistics. We could get driving-dependent exponents or a peak superimposed to the power law that might distort the estimate of the exponent value.

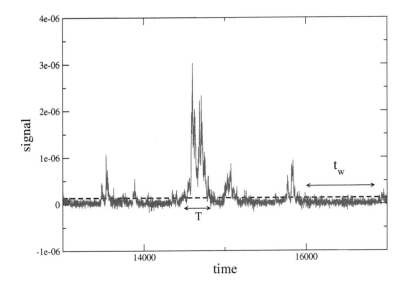

Fig. 1.9 A typical experimental avalanche signal, the Barkhausen noise in an amorphous ferromagnetic alloy, is composed by pulses of various durations T, separated by quiet time intervals of duration t_w. Due to the background instrumental noise it is necessary to set a threshold (dashed line) to define the avalanches. Source: Data courtesy of G. Durin.

When the driving rate is finite, it is also possible to study the waiting time t_w between different avalanches (Corral, 2006). This is also a statistical quantity, often distributed as

$$P(t_w) = t_w^\beta g_w(t_w/t^*), \tag{1.29}$$

defining another exponent β, a scaling function $g_w(x)$ and a cutoff t^* which is expected to decrease with the driving rate .

1.7 Background noise

So far we have considered the statistics in simple models where the avalanche events unambigously defined. This is not normally the case in experiments, where avalanches have to be identified from a signal that contains some level of backgound noise due to the instrumentation. This is shown by an experimental example of an avalanche signal reported in Fig. 1.9. In this case the signal $V(t)$ represents the voltage (Barkhausen) noise recorded by a coil around a ferromagnetic material during the magnetization process. The pulses are due to sudden avalanche-like changes in the magnetic flux across the sample. We will discuss this particular phenomenon in more detail in Chapter 8, using it here just as an illustrative example of the role of background noise.

Due to the background intstrumental noise, to properly identify the avalanches, we should set a threshold V_{th} and record the activity as anything exceeding the threshold (Janićević *et al.*, 2016). Once this is done, we can define for each pulse a duration T and the avalanche size s as the area under each pulse. There is some arbitrariness in the choice of the threshold and the results could in principle depend on it. This problem has been analyzed in Laurson *et al.* (2009) for the case of the Manna sandpile model discussed above. The model was driven with a small but non-vanishing rate h and the avalanche distributions were computed employing different thresholds. It was found that while for low threshold one finds the usual exponents, when the threshold is higher the exponent drifts slightly. In particular, the avalanche size distribution exponent changes from $\tau = 1.27$ to $\tau = 1.33$, while the avalanche duration distribution exponent changes from $\alpha = 1.5$ to $\alpha = 1.65$. An even more drastic behavior is observed for the waiting time distribution that crosses over from exponential to power law as a function of the threshold.

We should keep in mind this potential threshold dependence of the avalanche distribution when we analyze experimental data: To obtain reliable exponents it is necessary to have a signal with a high signal-to-noise ratio, so that the threshold can be kept small. An alternative that has been proposed in the context of Barkhausen noise is to apply a *Wiener filter* to the noise signal and then perform the usual analysis with a threshold set to zero (Papanikolaou *et al.*, 2011). The idea of the Wiener filter is to assume that the measured signal V_m can be written as

$$V_m(t) = \int dt' H(t - t')(V(t') + N(t')),\tag{1.30}$$

where $V(t)$ is the original signal, $N(t)$ is the background noise, and $H(t)$ is a response function. Working in Fourier space, given estimates for the response function $\hat{H}(\omega)$, the noise power spectrum $F_N(\omega)$, and a theoretical prediction for the power spectrum of the original signal $F_V(\omega)$, the filtered data can be obtained as the inverse Fourier transform of

$$\hat{V}(\omega) = \frac{\hat{V}_m(\omega) F_V(\omega)}{\hat{H}(\omega)(F_V(\omega) + F_N(\omega))}.\tag{1.31}$$

If we have good estimates for the background noise spectrum and knowledge about the expected spectrum of the original signal, we can obtain a filtered signal where the effect of the background noise is suppressed.

2

Sandpile Models

In this chapter we discuss deterministic and stochastic sandpile models. The relative simplicity of sandpile models has stimulated a series of theoretical approaches and we review here the most successful of these. As in standard critical phenomena, the simplest qualitative approach is provided by mean-field theory. A particularly instructive analogy in this respect is provided by branching processes, which can be seen as a general mean-field description of avalanche propagation. In the context of sandpile models, branching processes can also be used to describe the role of boundary dissipation on self-organization. While mean-field theory is general but approximate, some properties of sandpile models can be obtained exactly thanks to the Abelian properties of some sandpile models. Finally, we discuss field theory approaches to sandpile models and the relation with absorbing-state phase transitions.

2.1 Deterministic and stochastic sandpile models

As already discussed in Chapter 1, sandpile models are cellular automata (CA) where a number of grains z_i reside on a d-dimensional lattice. In the literature, the variable z_i is sometimes also referred to as 'energy". We will use this notation sometimes in the present chapter. In the original model introduced by Bak *et al.* (1987) grains are progressively added to randomly chosen lattice sites, until the number of grains on one site reaches a threshold z_c. When this happens the site relaxes

$$z_i \to z_i - z_c \tag{2.1}$$

and grains are transferred to the nearest neighbors sites j of the site i

$$z_j \to z_j + y_j \tag{2.2}$$

where y_j is the number of grains transferred to the site j. The relaxation of a site can induce nearest neighbor sites to relax on their turn, i.e. they exceed the threshold because of the grains received. New active sites can generate other relaxations and so on, eventually giving rise to an avalanche. When the avalanche is terminated new grains are randomly added to the lattice until a new avalanche is triggered. In the original sandpile model by Bak *et al.* 1987, known as the BTW model, $z_c = 2d$ and $y_j = 1$, while in the Manna model (Manna, 1991) (see Fig. 1.3) briefly discussed in Chapter 1, $z_c = 2$, independently of the dimensionality d, and two grains are distributed towards two randomly chosen nearest neighboring sites. Hence the main difference between the two models is that avalanches are deterministic in the BTW model and stochastic in the Manna model .

In both the BTW and Manna models , the dynamics is conserving the number of grains, since the number of transferred grains equals the number of grains lost by the relaxing site (i.e $z_c = \sum_j y_j$). Grain dissipation occurs only at the boundaries of the lattice, where grains can leave the system. The balance between the input and output of grains is the key ingredient for the sandpile to reach a steady-state characterized by an average constant "energy" $\langle z_i \rangle$ and by avalanche-like fluctuations. It is possible also to consider models with bulk dissipation (Manna *et al.*, 1990; Tadić *et al.*, 1992). For instance, one could introduce a probability $\epsilon > 0$ that a grain transferred by a relaxing site is lost. Dissipation breaks down the scale invariance of the avalanche distributions, effectively introducing a characteristic length in the system. Thus in the presence of bulk dissipation, the avalanche size distribution decays as

$$P(s) \sim s^{-\tau} f(s/s_c), \tag{2.3}$$

where the cutoff size scales as $s_c \sim \epsilon^{-1/\sigma}$. The lifetime distribution decays in a similar way as

$$P(T) \sim T^{-\alpha} g(T/T_c), \tag{2.4}$$

where T_c is a cutoff that scales as $T_c \sim \epsilon^{-\psi}$. The scaling functions f and g fall off exponentially for large arguments.

Over the past twenty years and more, a large numerical effort was devoted to characterize exponents and scaling functions for the different variants of the sandpile model (Manna, 1990; Grassberger and Manna, 1990; Chessa *et al.*, 1999; Lübeck, 2000; Lübeck, 2004), such as the stochastic Manna model (Manna, 1991) and the deterministic BTW model (Bak *et al.*, 1987). It was initially believed that the two models would belong to the same universality class (Chessa *et al.*, 1999). Numerical evidence was corroborated by a real space renormalization group analysis that indicated that all variants of the stochastic sandpile should fall into the same universality class which would also include deterministic models (Pietronero *et al.*, 1994; Vespignani *et al.*, 1995). A more refined and extensive numerical analysis finally showed quantitative differences in exponents and scaling functions for stochastic and deterministic sandpile models (Lübeck, 2000; Dickman and Campelo, 2003; Bonachela *et al.*, 2009). In particular, the avalanche distribution in the deterministic BTW model is not a simple power law, but the model nevertheless shows interesting mathematical features due to its Abelian property, which we discuss in the next section. On the other hand, stochastic sandpile models like the Manna model and other variants (Frette *et al.*, 1996; Paczuski and Boettcher, 1996; Amaral and Lauritsen, 1996) all fall into the same universality class, extremely well characterized numerically (Lübeck and Heger, 2003; Dickman and Campelo, 2003) as also predicted by field-theoretical analysis (Le Doussal and Wiese, 2015) discussed later in this chapter.

2.2   Branching Processes as a mean-field theory for avalanches

In analogy with equilibrium phase transitions, mean-field theory represents the simplest approach to reach a qualitative understanding of the system. Mean-field theories for sandpile models have been derived in different ways(Tang and Bak, 1988*a*; Tang and Bak, 1988*b*; Dhar and Majumdar, 1990; Vergeles *et al.*, 1997; Vespignani and

Zapperi, 1998), but the avalanche critical exponents (e.g. $\tau = 3/2$) always turn out to be the same. We can understand this noticing that an avalanche spreading can be described as a front composed of elements that could either trigger further spreading or die out. In a mean-field description, each element of the front will evolve independently with a spreading rate that only depends on the average properties of the system. Hence, we can consider an avalanche as a branching process , a well-studied mathematical model in which a set of independent elements increase their number or disappear with constant transition probabilities.

The relations between branching process es and SOC have been thoroughly investigated, and it has been proposed that the mean-field behavior of sandpile models can be described by a *critical* branching process (Alström, 1988; Zapperi *et al.*, 1995; Lauritsen *et al.*, 1996). To understand how a branching process becomes critical in the sandpile model, we discuss here the self-organized branching process (SOBP) (Zapperi *et al.*, 1995; Lauritsen *et al.*, 1996). The SOBP can be constructed as a mean-field theory for the Manna model with bulk dissipation in an infinite-dimensional lattice $d \to \infty$. In this model, a lattice site will topple upon addition of a grain only if it was already occupied. Or in other words, a toppling event will occur with probability p equal to the the density of occupied sites in the lattice. When a toppling occurs, two grains are transferred to two randomly selected neighboring sites with probability $1 - \epsilon$, where ϵ is the probability that a grain is lost due to dissipation. Since the lattice coordination number tends to infinity, the avalanche will never visit the same site twice. In other words, the toppling events are independent and are thus described by a branching process with branching ratio $R = 2p(1 - \epsilon)$. From the theory of branching process es (Harris, 1963), we know that there is a critical branching ratio, $R_c = 1$ such that for $R > R_c$ the probability to have an infinite avalanche is non-zero, while for $R < R_c$ all avalanches are finite. At the critical branching ratio, we expect avalanches to be power law distributed.

Sandpile models 'self-organize" to the critical point because they reach a steady-state where the grains added are balanced by the grains dissipated from the boundaries. We can introduce open boundary conditions in the mean-field theory by allowing for no more than n generations for each avalanche. Schematically, we can view the evolution of a single avalanche of size s as taking place on a tree of size $N = 2^{n+1} - 1$ (see Fig. 2.1). If the avalanche reaches the boundary of the tree, we count the number of active sites S_n, which in the sandpile language corresponds to the number of grains leaving the system. When grains leave the system the density p of grains in the lattice decreases, while if an avalanche stops before reaching the boundary, p increases.

To rewrite these arguments in a more formal way, let us consider the total number of grains $M(t)$ in the system after each avalanche. Here the time t is just counting the number of grains added between avalanches, which occurs instantaneously on this time scale. The number of grains evolve according to

$$M(t+1) = M(t) + 1 - S - K, \qquad (2.5)$$

where K is the number of particles lost by dissipation and S the number of particles that leave the system from the boundaries. Since $M(t) = Np$, we can rewrite Eq. 2.5 as an evolution equation for the parameter p

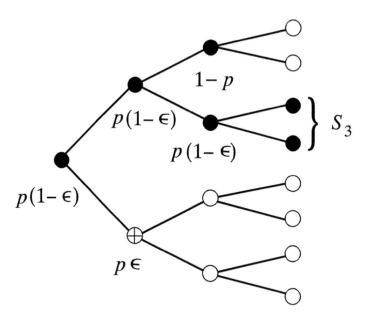

Fig. 2.1 Schematic drawing of an avalanche in a system with a maximum of $n = 3$ avalanche generations corresponding to $N = 2^{n+1} - 1 = 15$ sites. Each black site gives rise to two new black sites with probability $p(1-\epsilon)$, with probability $p\epsilon$ two particles are dissipated, and with probability $1 - p$ to two white sites. The black sites are part of an avalanche of size $s = 6$, whereas the active sites at the boundary yield $S_3(p, t) = 2$. The number of "stopped" sites $\mu = 2$, and there was one dissipation event such that $K = 2$. Source: Zapperi *et al.* (1995).

$$p(t + 1) = p(t) + \frac{1 - S(p, t) - K(p, t)}{N}. \tag{2.6}$$

In order to characterize the steady state of the SOBP model, we rewrite Eq. (2.6) in terms of the average values of S and K indicated by angular brackets. The average number of particles $\langle S_n \rangle$ leaving the system from the boundaries, in a system of n generations, is easily computed exploiting the recursive nature of the process

$$\langle S_n \rangle = (2p(1 - \epsilon))^n. \tag{2.7}$$

The evaluation of the average number of particles dissipated during an avalanche is a little more involved. We can first relate the average value of K to the average number of sites μ where an avalanche does not branch—either because of dissipation or because the site was empty (i.e., the avalanche stops). The two quantities K and μ are thus related according to

$$\langle K \rangle = 2 \langle \mu \rangle \frac{p\epsilon}{p\epsilon + 1 - p}. \tag{2.8}$$

The calculation of $\langle K \rangle$ then reduces to the calculation of $\langle \mu \rangle$. If we denote by S_m the number of active sites at generation m, then μ is given by

$$\mu = \sum_{m=0}^{n-1} \left(S_m - \frac{S_{m+1}}{2} \right) = \frac{1 + s - 2S_n}{2}, \tag{2.9}$$

where $s = \sum_{m=0}^{n} S_m$ is the total size of the avalanche.

The average value of s is obtained by summing the series:

$$\langle s \rangle = \sum_{m=0}^{n} \langle S_m \rangle = \frac{1 - (2p(1-\epsilon))^{n+1}}{1 - 2p(1-\epsilon)}. \tag{2.10}$$

We can combine these results and rewrite Eq. (2.6) in continuum notation as

$$\frac{dp}{dt} = \frac{\eta(p,t)}{N} + \frac{1}{N}\left(1 - \langle S_n \rangle - \frac{p\epsilon}{p\epsilon + 1 - p}\left(1 + \frac{1 - \langle S_{n+1} \rangle}{1 - 2p(1-\epsilon)} - 2\langle S_n \rangle\right)\right). \tag{2.11}$$

In Eq. (2.11), $\eta(p,t)$ describes the fluctuations around the average values of S and K. It can be shown numerically that the effect of the "noise" term η has a vanishingly small effect in the limit $N \to \infty$ (Lauritsen *et al.*, 1996). In absence of η, it is straightforward to study the fixed points of Eq. (2.11). The equation has only a single fixed point, $p^* = 1/2$ independently of the value of ϵ (Lauritsen *et al.*, 1996). By studying the stability of the fixed point, it is possible to show that it is attractive. Since the steady-state branching ratio is $R^* = (1-\epsilon)$, the SOBP model is *subcritical* as long as $\epsilon > 0$.

In Fig. 2.2, we show the value of p as a function of time for different values of the dissipation ϵ. Independently on the initial conditions, after a transient $p(t)$ reaches the steady-state described by the fixed point value $p^* = 1/2$ and fluctuates around it with short-range correlations. The fluctuations around the critical value decrease with the system size as $1/N$. Thus it follows that the distribution $\phi(p)$ approaches a delta function

$$\phi(p) \sim \delta(p - p^*) \tag{2.12}$$

in the limit $N \to \infty$.

In the limit where $n \gg 1$, it is possible to obtain exact predictions for the avalanche and lifetime distributions for any value of $\tilde{p} \equiv p(1-\epsilon)$. In particular, the critical branching process with $\tilde{p} = \tilde{p}_c = 1/2$ yields standard mean-field exponents $\tau = 3/2$ and $\alpha = 2$.

The quantities $P_n(s, \tilde{p})$ and $Q_n(S, \tilde{p})$ are defined as the probabilities of having an avalanche of size s and boundary size S in a system with n generations. The corresponding generating functions are defined by (Harris, 1963)

$$f_n(x, \tilde{p}) \equiv \sum_s P_n(s, \tilde{p}) x^s, \tag{2.13}$$

$$g_n(x, \tilde{p}) \equiv \sum_S Q_n(S, \tilde{p}) x^S. \tag{2.14}$$

From the hierarchical structure of the branching process , it is possible to write down recursion relations for $P_n(s, \tilde{p})$ and $Q_n(S, \tilde{p})$, from which we obtain (Harris, 1963)

$$f_{n+1}(x, \tilde{p}) = x\left[(1 - \tilde{p}) + \tilde{p}f_n^2(x, \tilde{p})\right], \tag{2.15}$$

$$g_{n+1}(x, \tilde{p}) = (1 - \tilde{p}) + \tilde{p}g_n^2(x, \tilde{p}), \tag{2.16}$$

where $f_0(x, \tilde{p}) = g_0(x, \tilde{p}) = x$.

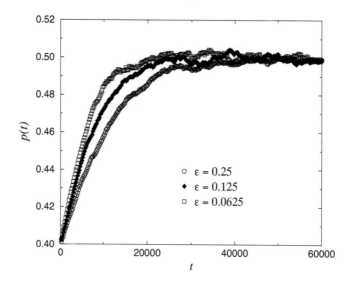

Fig. 2.2 The value of p as a function of time for a system with $n = 10$ generations. The curves refer to different values for ϵ. After a transient, the control parameter $p(t)$ reaches the steady-state value $p^* = 1/2$ and fluctuates around it with short-range correlations. Source: Lauritsen *et al.* (1996).

The solution of Eq. (2.15) in the limit $n \gg 1$ is given by

$$f(x, \tilde{p}) = \frac{1 - \sqrt{1 - 4x^2 \tilde{p}(1 - \tilde{p})}}{2x\tilde{p}}. \tag{2.17}$$

We expand Eq. (2.17) as a series in x, and by comparing with the definition (2.13), we obtain for sizes such that $1 \ll s < n$

$$P_n(s, \tilde{p}) = \frac{\sqrt{2(1 - \tilde{p})/\pi \tilde{p}}}{s^{3/2}} \exp\left(-s/s_c(\tilde{p})\right). \tag{2.18}$$

The cutoff $s_c(\tilde{p})$ is given by

$$s_c(\tilde{p}) = -\frac{2}{\ln 4\tilde{p}(1 - \tilde{p})}. \tag{2.19}$$

As $\tilde{p} \to 1/2$, $s_c(\tilde{p}) \to \infty$, thus showing explicitly that the critical value for the branching process is

$$\tilde{p}_c = 1/2. \tag{2.20}$$

Furthermore, we see that the mean-field exponent for the critical branching process is

$$\tau = 3/2. \tag{2.21}$$

We can expand $s_c(\tilde{p})$ in ϵ with the result

$$s_c(\epsilon) \sim \frac{2}{\epsilon^2}, \qquad \sigma = 1/2. \tag{2.22}$$

Eq. 2.22 shows the scaling of the characteristic avalanche size with the critical parameter ϵ.

The next step is to calculate the avalanche distribution $P(s)$ for the SOBP model. This can be calculated as the average value of $P_n(s, \tilde{p})$ with respect to the probability density $\phi(\tilde{p})$, that is according to the formula

$$P(s) = \int_0^1 d\tilde{p}\, \phi(\tilde{p})\, P_n(s, \tilde{p}). \tag{2.23}$$

The simulation results in Fig. 2.2 show that $\phi(\tilde{p})$ for $N \gg 1$ approaches the delta function $\delta(\tilde{p} - \tilde{p}^*)$ (cf. Eq. 2.12). Thus, for $1 \ll s < n$, we obtain the behavior

$$P(s) = \sqrt{\frac{2}{\pi}}\, \frac{1 + \epsilon + \cdots}{s^{-\tau}}\, \exp\left(-\frac{1}{2}\epsilon^2 s\right), \tag{2.24}$$

where $\tau = 3/2$ and the dependence of $\tilde{p}^*$ on ϵ has been made explicit.

These results are in perfect agreement with the simulation of the model shown in Fig. 2.3. The deviations from the power-law behavior (2.24) are due to the fact that Eq. (2.18) is only valid for $1 \ll s < n$. For small s we have the exact results $P_n(1, p) = 1 - p$, $P_n(3, p) = p(1 - p)^2$, $P_n(5, p) = 2p^2(1 - p)^3$, and so forth.

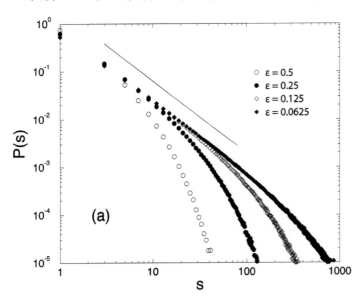

Fig. 2.3 Log-log plot of the avalanche distribution $P(s)$ for different levels of dissipation. A line with slope $\tau = 3/2$ is plotted for reference, and it describes the behavior of the data for intermediate s values, cf. Eq. 2.24. For large s, the distributions fall off exponentially. Source Lauritsen *et al.* (1996).

The avalanche lifetime distribution $P(T)$ is defined as the probability to obtain an avalanche which spans m generations; here, we identify m with the time T. It follows that

$$P(m, \tilde{p}) = Q_{m+1}(S = 0, \tilde{p}) - Q_m(S = 0, \tilde{p}). \tag{2.25}$$

For $\tilde{p} = 1/2$ we use the general result (Harris, 1963)

$$\frac{1}{1 - Q_m(S = 0, \tilde{p})} = 1 + m\tilde{p} + O(\ln m), \qquad m \gg 1, \tag{2.26}$$

and obtain

$$P(T, \tilde{p}) = \frac{\tilde{p}^{-1}}{t^2} \left(1 + O(\ln T/T) + \ldots\right). \tag{2.27}$$

Note the strong correction to scaling to $P(T)$.

For $\tilde{p} < 1/2$ we have (Harris, 1963)

$$1 - Q_m(S = 0, \tilde{p}) \sim c_1 (2\tilde{p})^m, \tag{2.28}$$

for $m \gg 1$, where $c_1 > 0$ is an unknown constant. This expression yields $P(T) \sim \epsilon \exp(-n\epsilon)$. We can combine the above results in the scaling form

$$P(T) \sim T^{-\alpha} \exp(-T/T_c), \tag{2.29}$$

where

$$T_c \sim \epsilon^{-\psi}, \qquad \psi = 1, \tag{2.30}$$

and $\alpha = 2$.

2.3 Abelian sandpiles

2.3.1 General formalism

The formalism of the Abelian sandpile model (ASM) was introduced by Dhar and collaborators in a series of papers (Dhar, 1990; Dhar, 1999; Dhar and Ramaswamy, 1989; Dhar and Majumdar, 1990; Majumdar and Dhar, 1992). The dynamics of a class of sandpile models can be characterized by an Abelian group, a property that can be exploited to obtain rigorous results and, in some special cases, exact values for critical exponents.

The ASM is defined by a set of rules that generalize the deterministic BTW model (Dhar, 1990). A set of N sites on a lattice is labeled with indices i running between 1 and N and we then assign to each site an integer variable z_i. At each time step, a site i is chosen at random and its variable is increased: $z_i \rightarrow z_i + 1$. We then specify a set of thresholds z_{ic} for $(i = 1...N)$ and an $N \times N$ integer toppling matrix Δ_{ij}. If $z_i > z_{ic}$ the site relaxes and grains are transferred to the other sites according to the toppling matrix:

$$z_j \rightarrow z_j - \Delta_{ij}. \tag{2.31}$$

The same procedure is iterated until no unstable sites are present. The boundary conditions are open so that the energy can leave the system. To this end, it is useful for this to define an extra site (the *sink*) connected with all the boundary sites. In general, the matrix Δ_{ij} should satisfy the following conditions:

$$\Delta_{ii} > 0, \text{ for } i = 1, ...N, \tag{2.32}$$

$$\Delta_{ij} < 0, \text{ for } i \neq j, \tag{2.33}$$

and

$$\sum_{i=1}^{N} \Delta_{ij} \geq 0, \text{ for } i = 1, ...N. \tag{2.34}$$

In the case of the BTW model, Δ_{ij} is a symmetric matrix with diagonal elements $\Delta_{ii} = 4$. The elements connecting nearest neighbors are equal to $\Delta_{ij} = -1$ and all the elements are zero.

If we consider the case in which $z_{ic} = \Delta_{ii}$, a configuration $\{z_i\}$ is stable if $z_i < \Delta_{ii}$. In the space of stable configurations, we define the operators a_i $(i = 1...N)$ that perform the addition of a grain in i. If C is a stable configuration, $a_i C$ is the configuration obtained by adding a grain in i and letting the system relax. It is possible to show that the operators a_i commute with each other:

$$[a_i, a_j] = 0, \text{ for all } i, j. \tag{2.35}$$

This Abelian property turns out to be crucial for deriving exact results (Dhar, 1999). Dhar divides the set of stable configurations into two classes (Dhar, 1990): recurrent and transient. A configuration C is defined to be recurrent, if N positive integer numbers m_i exist such that

$$a_i^{m_i} C = C, \text{ for all } i. \tag{2.36}$$

Due to the Abelian property, the set of recurrent configurations is closed under application of a_i. From this observation it follows that it is possible to identify the set of recurrent configurations with the self-organized critical steady state, which is thus an attractor of the dynamics. Furthermore, it can also be shown that all the configurations in this state have the same probability of occurrence (Dhar, 1990).

2.3.2 The burning algorithm and the spanning tree

To characterize the steady state it is useful to define a forbidden sub-configuration (FSC) as any set F of r sites for which

$$z_j \leq \sum_{i \in F, i \neq j} (-\Delta_{ij}), \text{ for all } j \in F. \tag{2.37}$$

A configuration that contains no FSC is called an allowed configuration. It can be rigorously proven that the set of stable allowed configurations is the same as the set of recurrent configurations (Dhar, 1990; Gabrielov, 1993).

The result discussed above can be used to introduce an algorithm to distinguish between recurrent and transient configurations (Dhar, 1990). Consider a configuration T: one can check if $T = F$ by testing the inequality 2.37. If the hypothesis is not true, one can remove (or *burn*) from the set all the sites for which the inequality is not satisfied and see if the remaining sub-configuration is forbidden. Finally T will become empty and the configuration is therefore recurrent, or some FSC will be found showing that the configuration is transient. This algorithm is equivalent to the *toppling from sink* algorithm: Given a configuration of the ASM one can increase by one all the boundary sites and then let the system relax. The configuration is allowed if each site of the lattice topples once and only once.

So far we have not considered the order in which sites are burned. It is instructive, however, to consider a time of burning (Majumdar and Dhar, 1992). A site i is burned at time t if at time $t-1$ z_i was larger than the number of bonds connecting i to other unburnt sites. Drawing the bond between successively burnt sites, we obtain a spanning tree covering the lattice. This implies that the recurrent configurations of the ASM are in a one-to-one correspondence with spanning trees. It is known that the spanning tree problem is equivalent to the q-state Potts model in the limit $q \to 0$. The Potts model is a well-studied generalization of the Ising model where the lattice variables have q possible states $s_i = 1, ...q$ and the energy is given by

$$E = -J \sum_{ij} \delta_{s_i,s_j}. \tag{2.38}$$

It has been argued that it would be possible to apply to the ASM results known for the Potts model. If one assumes that the burning process would correspond to the spreading of an avalanche, one could conclude that the dynamical exponent for the ASM has the value $z = 5/4$ found in the Potts model. In fact this value turns out to be close to those obtained in computer simulations of sandpile models. A more precise connection between the avalanche dynamics and the spanning trees will be discussed in the next section.

2.3.3 Waves of toppling

Ivashkevich *et al.* described a decomposition of the avalanche in a series of waves of toppling which in their turn can be described as spanning trees (Ivashkevich *et al.*, 1994). Consider an avalanche starting from a site j within the lattice. We let the avalanche spread until site j becomes unstable for the second time. At this point, we freeze site j until all the other sites relax. In this way, we can identify the first *wave* of the avalanche. When the first wave is terminated, we re-activate site j, allowing the avalanche to spread, but preventing site j from toppling a third time. In this way an avalanche is decomposed in a set of waves $W_1, W_2....W_n$, composed by sites that topple only once. It is possible to associate to each wave a two-rooted spanning tree covering the lattice, composed by two disconnected trees which span the lattice and have the *sink* and the site j as their roots.

This mapping allows one to link the avalanche properties with those of the lattice Green function for the Poisson equation. In particular, Ivashkevich *et al.* demonstrate the following equation:

$$G_{ij} = N^{(i,j)}/N_t \tag{2.39}$$

where $G_{ij} = (\Delta^{-1})_{ij}$ is the lattice Green function; $N^{(i,j)}$ is the total number of two-rooted spanning trees, having roots i_0 and j, such that i and j belong to the same sub-tree; and N_t is the total number of spanning trees in the lattice. From the asymptotic properties of the lattice Green function, it is possible to derive the probability distribution of the waves. In particular, for a wave starting inside the lattice it is possible to show that

$$P(w) \sim 1/w, \tag{2.40}$$

where $P(w)$ is the probability of having a wave of size w. It is not possible to obtain exactly the distribution of avalanche sizes directly from the wave distribution, except in the case of avalanches starting from the boundaries, since these avalanches consist of a single wave. In this case, one obtains a boundary critical exponent for the size distribution given by $\tau_b = 3/2$. The result is generalized to boundaries forming an arbitrary angle θ where one obtains $\tau_b = 1 + \frac{\pi}{2\theta}$.

2.3.4 Other results

ASM, in view of their analytical tractability have stimulated a number of interesting analyses leading to exact results for various models variants. We do not attempt to discuss this whole body of work here, but we quote some of the results, referring to the literature for the derivations and to Dhar (2006) for a more comprehensive review.

1. *Height distributions.* In two dimensions, the probability of having $z_i = 1$ in the critical state is calculated as $P(1) = (2/\pi^2) - (4/\pi^3)$ (Majumdar and Dhar, 1991). The result was also generalized to $P(z)$ with $z > 1$ (Priezzhev, 1994). The probability that two sites separated by distance r will both have $z = 1$ is $P(1,1) = P(1)^2(1 - 1/2r^4 + \ldots)$ (Majumdar and Dhar, 1991). Note that this scaling is similar to that of the energy correlation function in the $q \to 0$ Potts model.

2. *Directed models.* The directed ASM can be solved exactly (Dhar and Ramaswamy, 1989). The exponents of the distributions of avalanche durations and sizes are given by $\alpha = 3/2$ and $\tau = 4/3$ in two dimensions, and $\alpha = 2$ and $\tau = 3/2$ for $d \geq 3$. Therefore $d_c = 3$ is the upper critical dimension for the directed ASM. It has also been shown that avalanches are compact and leave no holes (i.e. $D = d$). Small changes in the model do not change its universality class, which has been shown to be the same for the triangular lattice and for a partially directed variant of the model.

3. *Bethe lattice* The Bethe lattice is a hierarchical tree where no loops are present and therefore the behavior of the model is expected to be mean field. By writing recursion relations it is possible to compute single site probabilities $P(z_i = j), j = 1, .., ., z_c$ and pair distribution functions (Dhar and Majumdar, 1990). The correlations between sites decay exponentially with distance. The avalanche distribution decays with the mean-field exponent $\tau = 3/2$.

4. *One-dimensional models.* Considerable effort has been put into solving one-dimensional ASMs. Most of these models, however, do not exhibit scaling behavior (Ruelle and Sen, 1992; Ali and Dhar, 1995).

2.4 Sandpiles and absorbing-state phase transitions

The steady state of sandpile models is produced by the balance between the input and loss of grains. If grains are added at rate h and dissipated at rate ϵ, the steady state is characterized by a density of active sites (i.e. those sites for which $z \geq z_c$) $\rho_a = h/\epsilon$ (Vespignani and Zapperi, 1997). The systems is critical in the limit of vanishing activity and dissipation $h \to 0$ and $\epsilon \to 0$, with $h/\epsilon \to 0$ (Vespignani and Zapperi, 1998; Dickman *et al.*, 1998). Setting these two parameters to zero represents a sort of fine-tuning, albeit peculiar, since the tuning is only reached at a limit. When the driving rate is non-vanishing but small, the system jumps among stable configurations through avalanches. After each avalanche, the sandpile reaches a stable configuration and would remain there if not perturbed by further input of grains. We may call these static configurations *absorbing states* in analogy with other lattice models, like the contact process (Grassberger and de la Torre, 1979), undergoing absorbing state phase transition (Hinrichsen, 2000). In these models, depending on the value of some parameters, the system ends up either in an active or in an absorbing phase, with a phase transition between the two.

Absorbing phase transitions have been widely studied as examples of simple non-equilibrium critical phenomena (Hinrichsen, 2000). The analogy between sandpiles and other absorbing-state phase transitions can be clarified considering a version of the sandpile model where the number of grains is strictly conserved (i.g. $h = 0$, $\epsilon = 0$ and periodic boundary conditions), a model often denoted as the *fixed-energy sandpile* (Vespignani *et al.*, 1998; Vespignani *et al.*, 2000). Since the number of grains is conserved, the grain density $\zeta = \sum_i z_i/L^d$ can be used as the control parameter for the model. If the density ζ is large enough, the system reaches a stationary active state, while for small ζ the system relaxes into one of the many possible absorbing states. Simulations show that the two phases are separated by a second-order phase transition at a critical value $\zeta = \zeta_c$ (Dickman *et al.*, 2000).

Self-organized criticality in sandpile models is easier to understand from the perspective of absorbing-state phase transitions. In the driven-dissipative sandpile model, grains are only added when no activity is present (i.e. for $\zeta < \zeta_c$), while conversely dissipation can only take place when there *is* activity (i.e. for $\zeta > \zeta_c$). In other words, $d\zeta/dt > 0$ for $\zeta < \zeta_c$, and $d\zeta/dt < 0$ for $\zeta > \zeta_c$, which implies that the system is naturally driven towards the critical point $\zeta = \zeta_c$.

The analogy with absorbing-state phase transitions was also useful to derive a field theory for stochastic sandpiles that could help compute their scaling exponents and identify their universality class. Contrary to many other models whose absorbing-state phase transition is described by the directed percolation universality class (Hinrichsen, 2000), stochastic sandpiles define a different universality class. The key reason for this is the presence of a grain conservation law in sandpile dynamics that is not present in the other models.

Here we follow Vespignani *et al.* to describe general continuum equations for sandpiles based on general symmetry considerations. The equations describe the evolution of the order parameter field, the local density of active sites, $\rho_a(\mathbf{x})$. When $\rho_a(\mathbf{x}) = 0$ for all $\mathbf{x}$, the system is an absorbing configuration and in absence of other perturbations it will remain there. The order parameter field $\rho_a(\mathbf{x})$ should be coupled to the local grain density, $\zeta(\mathbf{x}, t)$, which influences the generation of new active sites and it is a *conserved* field in the fixed energy sandpile. The most general evolution equation compatible with local conservation is

$$\frac{\partial \zeta(\mathbf{x}, t)}{\partial t} = \nabla^2(f_\zeta[\{\rho_a\}, \{\zeta\}]) + \nabla \cdot [g_\zeta(\{\rho_a\}, \{\zeta\}) \vec{\eta}(\mathbf{x}, t)], \tag{2.41}$$

where f_ζ and g_ζ are functionals of ρ_a and ζ. Conservation is enforced by the Laplacian term and by d-component vectorial conserving noise $\vec{\eta}$ which is standard in non-equilibrium critical phenomena (Bray, 2002). The evolution equation for the order parameter field can be written in full generality as

$$\frac{\partial \rho_a(\mathbf{x}, t)}{\partial t} = f_a(\{\rho_a\}, \{\zeta\}) + g_a(\{\rho_a\}, \{\zeta\})\eta(\mathbf{x}, t), \tag{2.42}$$

where f_a and g_a are functionals of ρ_a and ζ and $\eta(\mathbf{x}, t)$ is an uncorrelated non-conserving Gaussian noise.

One can write explicit forms of functionals based on symmetry considerations and an expansion to the lowest relevant order in the fields (see Vespignani *et al.* (2000) for more details). Under these conditions the activity equation takes the form

$$\frac{\partial \rho_a(\mathbf{x}, t)}{\partial t} = D_a \nabla^2 \rho_a(\mathbf{x}, t) - r\rho_a(\mathbf{x}, t) - b\rho_a^2(\mathbf{x}, t)$$
$$+ \mu\rho_a(\mathbf{x}, t)\Delta\zeta(\mathbf{x}, t) + \eta_a(\mathbf{x}, t), \tag{2.43}$$

where $\eta_a = \rho_a^{1/2}\eta$ and $\Delta\zeta = \zeta(\mathbf{x}, t) - \zeta_0$. Here the average energy density ζ_0 plays the role of control parameter.

To the lowest order, the evolution of $\Delta\zeta(\mathbf{x}, t)$ can be integrated to yield

$$\Delta\zeta(\mathbf{x}, t) = \Delta\zeta(\mathbf{x}, 0) + \int_0^t dt' \left[D_\zeta \nabla^2 \rho_a(\mathbf{x}, t') + \nabla \cdot \left(\sqrt{\rho_a(\mathbf{x}, t')} \vec{\eta} \right) \right]. \tag{2.44}$$

Substituting Eq. 2.44 into Eq. 2.43, we finally reach a Langevin equation for fixed-energy sandpiles (Vespignani *et al.*, 1998; Vespignani *et al.*, 2000),

$$\frac{\partial \rho_a(\mathbf{x}, t)}{\partial t} = D_a \nabla^2 \rho_a(\mathbf{x}, t) - r(\mathbf{x})\rho_a(\mathbf{x}, t) - b\rho_a^2(\mathbf{x}, t)$$
$$+ w\rho_a(\mathbf{x}, t) \int_0^t dt' \nabla^2 \rho_a(\mathbf{x}, t') + \sqrt{\rho_a}\eta(\mathbf{x}, t). \tag{2.45}$$

Eq. 2.45 should describe a broad universality class encompassing all the absorbing state phase transition with a conserved field which would include all stochastic sandpiles (Rossi *et al.*, 2000; Lübeck and Heger, 2003; Ódor, 2004). It is straightforward to show by standard power-counting analysis that the upper critical dimension of this theory is $d_c = 4$ (Vespignani *et al.*, 2000). Computing the critical exponents for $d < d_c$ has been a challenge for more than a decade.

The issue was finally solved in 2015 by Pierre Le Doussal and Kay Wiese who showed that the field theory could be mapped exactly to the continuum theory for the depinning of an elastic interface moving in a random medium (Le Doussal and Wiese, 2015)— a topic that will be addressed in detail in Chapter 4. The idea that sandpiles and depinning interfaces were somehow related has been discussed for quite some time and was based on mappings between variants of the sandpile model and discretized equations for interface depinning (Paczuski and Boettcher, 1996; Alava and Lauritsen, 2001; Alava, 2002; Alava and Munoz, 2002). Furthermore, a numerical analysis of cusp singularities of the effective coarse-grained disorder correlator revealed remarkable similarities between systems with multiple absorbing states and depinning interfaces (Bonachela *et al.*, 2009). Finally, thanks to Le Doussal and Wiese the mapping between sandpiles and interface depinning was made more rigorous (Le Doussal and Wiese, 2015). Their work provided a complete characterization of the universality class of stochastic sandpiles. As we will show in Chapter 4, the critical exponents of interface depinning are known with great precision thanks to the functional renormalization group theory developed in the early '90s by the groups of Thomas Nattermann(Nattermann *et al.*, 1992) and Daniel Fisher (Narayan and Fisher, 1993) and later perfected by Le Doussal, Wiese, and co-workers (Chauve *et al.*, 2001; Le Doussal *et al.*, 2009; Le Doussal and Wiese, 2013; Dobrinevski *et al.*, 2015).

3
Avalanches in disordered media

Several avalanche phenomena occur in disordered media where randomness is frozen or "quenched" and does not evolve on the timescale of the avalanches. The simple examples are provided by percolation and its dynamic counterpart, invasion percolation. Furthermore, a wide class of driven disordered systems displays athermal disorder-induced phase transitions characterized by avalanche dynamics. The prototype model for this behavior is the random-field Ising model that, at the critical point, displays power-law distributed-avalanche distributions that can be understood analytically. The model is particularly interesting because it is the prototype of many problems ruled by the competition between nucleation and growth of domains in a disordered landscape.

3.1 Percolation

Many avalanche phenomena in nature are rooted in *quenched disorder*, which describes the presence degrees of freedom that are essentially frozen on the rapid time scale of avalanche propagation. The dynamics of a rapidly evolving activity field over a static grain landscape is underlying the physics of sandpile models. The presence of a hidden quenched disorder becomes fully apparent considering that those models are mapped into interfaces moving on random media (Le Doussal and Wiese, 2015). It is therefore important to study avalanche dynamics in systems with quenched disorder.

The simplest lattice model with quenched disorder is percolation, where the sites (or the bonds) of a d-dimensional lattice are randomly occupied with a probability p (for a review see Stauffer and Aharony (1994). As the name suggests, the model is inspired by the percolation of a fluid through a random porous medium. Pores can allow locally the passage of the fluid, and the question is whether the fluid entering from one end of the medium is able to percolate to the other side. The answer to this question boils down to the presence of a path connecting the two ends. Percolation is relevant for a wide variety of problems involving transport in random media, including fluid flow in rocks and soil (Berkowitz and Ewing, 1998), coffee extraction (Blumberg *et al.*, 2010), and conduction in disordered solids (Kirkpatrick, 1973).

Coming back to the lattice model, we distinguish between bond and site percolation depending on which element can be occupied with probability p. In both cases, there is a critical value p_c above which there is a path connecting the two ends of the lattice. For $p < p_c$, occupied regions form a set of disconnected clusters whose average size increases with p and in the limit of an infinite system diverges when $p \to p_c$ (see Fig. 3.1).

The percolation transition has all the features of second-order phase transitions, namely a critical point, scaling behavior, and universal critical exponents. Universality

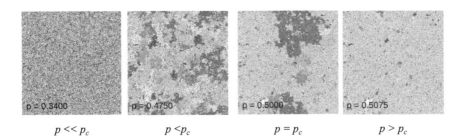

$p \ll p_c$ $p < p_c$ $p = p_c$ $p > p_c$

Fig. 3.1 Simulations of bond percolation on a square lattice for different values of p. Clusters are shown in different shades of grey. At $p = p_c = 1/2$ a spanning cluster appears. Adapted from images by Thierry Dugnolle (CC-BY-SA-4.0 licence).

here means that the value of the exponent does not depend of the type of percolation (site or bond) or the type of lattice (square or triangular) but only on the dimensionality of the lattice.

At the percolation point $p = p_c$, the system is scale invariant and can be described by power-law relations. In particular, the percolation cluster is a fractal (Fig. 3.2) so its mass M scales with the lattice size L as

$$M \sim L_f^d, \tag{3.1}$$

where the fractal dimension is known exactly in $d = 3$ where it is$d_f = 91/48 \simeq 1.89$ and numerically in $d = 3$ where it is equal to $d_f \simeq 2.5$.

At $p = p_c$, the probability distribution of clusters of size c is also a power law scaling as (Hoshen *et al.*, 1979)

$$P(c) \sim c^{-\tau_c}, \tag{3.2}$$

with $\tau_c = 187/91$ in $d = 2$ and $\tau \simeq 2.19$ in $d = 3$. The approach to the percolation point for $p < p_c$ is also described by scaling laws. For instance the average cluster size scales as

$$\langle c \rangle \sim (p_c - p)^{-\gamma} \tag{3.3}$$

with $\gamma = 43/18$ in $d = 2$ and $\gamma \simeq 1.8$ in $d = 3$. The largest cluster size s_0 scales as

$$c_0 \sim (p_c - p)^{-1/over\sigma} \tag{3.4}$$

with $\sigma = 36/91$ in $d = 2$ and $\sigma \simeq 0.45$ in $d = 3$. We can also define the characteristic length of the cluster ξ scaling as

$$\xi \sim (p_c - p)^{-\nu} \tag{3.5}$$

with $\nu = 4/3$ in $d = 2$ and $\nu \simeq 0.88$ in nu$d = 3$. Combining Eqs. 3.5, 3.4, and 3.1, we can readily derive the scaling relation $1/\sigma = d_f \nu$.

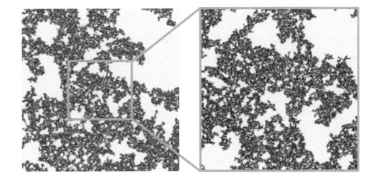

Fig. 3.2 Simulation of a critical percolation cluster for bond percolation on a square lattice at $p = p_c = 1/2$. When a part of the cluster is magnified it is statistically indistinguishable from the whole cluster as expected for a scale invariant fractal structure. Source: Adapted from images by Thierry Dugnolle (CC-BY-SA-4.0 licences).

3.2 Invasion percolation

The reader should have noticed the similarity between the scaling exponents describing clusters in percolation and those introduced in section 1.3 to describe avalanche statistics in crackling noise. Percolation clusters, however, are not avalanches but static geometrical objects. To make the connection between avalanches and percolation clusters, we need to consider dynamic models of percolation such as invasion percolation (Wilkinson and Willemsen, 1983).

In invasion percolation, the sites (or bonds) of a d-dimensional lattice are endowed with random thresholds x_i, uniformly distributed in $[0, 1]$. To make things concrete, we could interpret x_i as the value of the fluid pressure needed by the invading fluid to enter a pore i occupied by a defending fluid. The invasion process starts from a single site j at the edge of the lattice and proceeds through the site with the lowest value of x_i among the nearest neighbors of j. Since we are describing the motion of a fluid, one should distinguish between two scenarios depending on the compressibility of the defending fluid. If the defending fluid is compressible, the invading fluid can potentially enter any available region. On the other hand, if the defending fluid is incompressible it becomes trapped when a region is surrounded by the invading fluid. As a consequence of this the invading fluid can not enter in that region anymore. The two models are called nontrapping invasion percolation (NTIP) and trapping invasion percolation (TIP), respectively (Fig 3.3).

As the invasion process proceeds, the invaded area follows a fractal pattern (see Fig. 3.4) with a fractal dimension that, in the case of NTIP, is consistent to the one describing the spanning cluster in standard random percolation, that is $d_f \simeq 1.89$ in

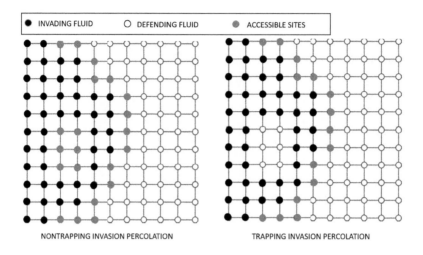

Fig. 3.3 The rules of invasion percolation depend on the compressibility of the defending fluid. If the defending fluid is incompressible, regions enclosed by the invading fluid become inaccessible (trapping).

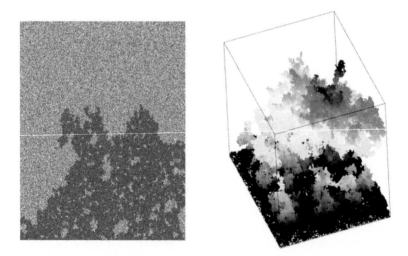

Fig. 3.4 The cluster of invasion percolation with trapping in $d = 2$ and $d = 3$. Images from Nooruddin and Blunt 2018, CC-4.0 licence.

$d = 2$ and $d_f \simeq 2.5$ in $d = 3$ (Sheppard *et al.*, 1999). For TIP, the fractal dimension of the invading cluster in $d = 2$ is $d_f \simeq 1.8$, smaller than the expected for random percolation, while no significant differences are found in $d = 3$.

   While invasion percolation dynamics occurs one site or bond a time, it is still possible to define avalanches by ordering the random threshold x_i by the time of invasion t (Roux and Guyon, 1989; Furuberg *et al.*, 1988; Maslov, 1995; Cafiero *et al.*, 1996). The sequence $x_i(t)$ is randomly fluctuating, as illustrated in Fig. 3.5. Since invasion percolation follows the path of lowest local threshold, only low-threshold sites/bonds are initially invaded, but then larger and larger thresholds are invaded. Starting from a site $i0$ invaded at time t_0, the avalanche is defined considering the sequence of invaded sites/bonds for which $x_i(t) < x_{i0}(t_0)$. When $x_i(t) > x_{i0}(t_0)$, the current avalanche ends and a new avalanche starts. Numerical simulations in $d = 2$ show that with this definition avalanches in invasion percolation are distributed as a power law with exponents $\tau = 1.6$ for TIP and $\tau = 1.53$ for NTIP (Maslov, 1995). For NTIP it is also possible to relate the avalanche exponent with other standard percolation exponents using a scaling relation (Roux and Guyon, 1989).

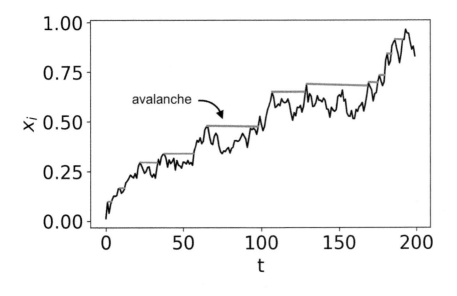

Fig. 3.5 To define avalanches in invasion percolation we plot the random threshold x_i ordered by the time of invasion t. The avalanche starting at time t_0 is defined counting the number of sites invaded for which $x_i(t) < x_{i0}(t_0)$. Avalanches are indicated in gray in the plot.

3.3 Avalanches in the random-field Ising model

Sethna *et al.* have proposed the driven random-field Ising model (RFIM) as a prototype for avalanches in disordered media undergoing an athermal first-order phase transition. The model was extensively studied analytically and numerically, with its scaling behavior eventually associated with a disorder-controlled critical point (Dahmen and Sethna, 1993; Perkovic *et al.*, 1995; Dahmen and Sethna, 1996; Perkovic *et al.*, 1999; Sethna *et al.*, 2001). In the RFIM, a set of spins $s_i = \pm 1$ is assigned to each site i of a d-dimensional lattice. The spins interact with their nearest neighbors' spins through a coupling constant J and with an externally applied field H. Disorder is modeled by associating each lattice site with a random field h_i taken from a Gaussian probability distribution with variance R, $\rho(h) = \exp(-h^2/2R^2)/\sqrt{2\pi}R$. The Hamiltonian thus reads

$$\mathcal{H} = -\sum_{\langle i,j\rangle} J s_i s_j - \sum_i (H + h_i) s_i, \tag{3.6}$$

where the first sum is restricted to nearest-neighbors pairs. In the zero-temperature dynamic used by Sethna *et al.* (1993), the spins align with the local field $f_i \equiv -\partial\mathcal{H}/\partial s_i$ (Bertotti and Pasquale, 1990). Thus each spin evolves according to

$$s_i = \text{sign}(f_i) = \text{sign}(J \sum_j s_j + h_i + H). \tag{3.7}$$

This dynamic obeys return-point memory (Sethna *et al.*, 1993): If the field is increased adiabatically, the magnetization only depends on the state in which the field was last reversed. This property has been exploited in $d = 1$ (Shukla, 2000; Dante *et al.*, 2002), and in the Bethe lattice (Shukla, 2001; Colaiori *et al.*, 2002) to obtain exactly the shape of the magnetization curve for various field histories. Here we consider the simplest case in which the external field is ramped adiabatically from $-\infty$ to ∞. Notice that this prescription is equivalent to the Glauber dynamics at temperature T and with an external field ramped at rate $\dot{H}$ taking the limit $T \to 0$ first and then $\dot{H} \to 0$.

At the beginning of the ramp, all spin points in the downward direction until the local field of a spin changes sign, causing the spin to flip. This event in turn increases the local field of the neighboring spins by $2J$ which could then also flip, eventually triggering an avalanche. For small values of the disorder R, the first spin flips are likely to generate a substantial avalanche of size comparable to the system size, leading to a discontinuous magnetization change. In the opposite limit of a large disorder, the spin reverses smoothly and only small avalanches are observed. In fact, these two regimes are separated by a critical disorder R_c, where the avalanches are distributed as a power law. The behavior of the model close to R_c is very similar to that of equilibrium systems at the critical points and can thus be studied by standard methods, like mean-field theory and the renormalization group. It is also possible to perform extensive numerical simulations to extract the scaling behavior. Notice that in this model we can measure the full spectrum of avalanche distributions and exponents discussed in previous sections. Particularly, the avalanche size S is defined as the number of spins reversed after a single spin has been flipped, or in other words, by the change of magnetization. The avalanche duration T is the number of updating steps, in parallel dynamics, taking place during the avalanche.

3.4   Critical behavior of the driven RFIM

A qualitative picture of the behavior of the RFIM can be obtained by mean-field theory (Sethna *et al.*, 1993; Dahmen and Sethna, 1996). To this end, we consider Eq. 3.6, extending the sums to all the pairs of sites, and obtain

$$E = -\sum_i (JM + H + h_i) s_i, \tag{3.8}$$

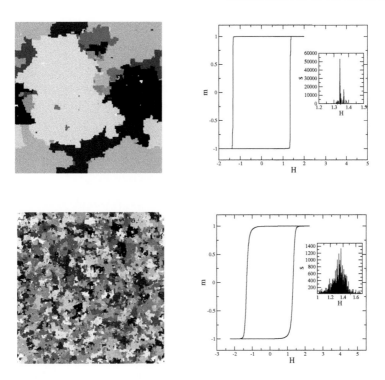

Fig. 3.6 Avalanches in the two-dimenensional RFIM. The graphs on the left display the spatial arrangement of the avalanches (each tone of gray represents one avalanche). The right panels display the hysteresis loop: the magnetization per spin as a function of the applied field. In the insets, we report the corresponding avalanche sizes. In the top panels, we report the simulations for $R/J = 0.8$, while in the bottom panels $R/J = 1.2$. Here the lattice size is $L = 400$. Notice the lartge qualitative differences in the avalanches due to a relatively small variation of R/J.

where $M = \sum_i s_i$. The magnetization can then be obtained self-consistently as

$$M = 1 - 2P(JM + H + h_i < 0) = 1 - 2\int_{-\infty}^{-JM-H} \rho(h)dh. \qquad (3.9)$$

This equation has a single-valued solution for $R > R_c = \sqrt{(2/\pi)}J$, while the solution becomes multivalued for $R < R_c$. In the second case, we have a hysteresis loop with a jump at H_c and a diverging slope dM/dH. For high disorder, there is no hysteresis, but this is an artifact of mean-field theory that is not observed in low dimensions. At the critical point $R = R_c$ the magnetization has a diverging slope, but no finite jump. In analogy with critical phenomena it is possible to obtain scaling laws describing the singularities occurring at the transition. In particular the magnetization obeys the scaling law

$$M = |r|^\beta g_\pm(h/|r|^{\beta\delta}), \tag{3.10}$$

where $r \equiv (R - R_c)/R_c$ and $h \equiv (H - H_c)/H_c$ are the reduced control parameters, and $\beta = 1/2$, $\delta = 3$. The scaling function $g_\pm$ is obtained as the root of a cubic equation.

The distribution of avalanche sizes near the critical point can be computed analytically in the framework of mean-field theory. The result can be summarized in the scaling form

$$P(S, r, h) = S^{-\tau} \mathcal{P}_\pm(S/|r|^{-1/\sigma}, h/|r|^{\beta\delta}), \tag{3.11}$$

where the exponents take the mean-field values $\tau = 3/2$, $\sigma = 1/2$. The scaling function $\mathcal{P}_\pm$ can be evaluated exactly in terms of $g_\pm$ and is given by

$$\mathcal{P}_\pm(x, y) = (1/\sqrt{2\pi}) \exp(-x(1 \mp \pi g_\pm(y)^2/4)^2/2). \tag{3.12}$$

In addition to the size distribution, one can also define the distribution of durations T, which obeys an analogous scaling form

$$D(T, r, h) = T^{-\alpha} \mathcal{D}_\pm(T/|r|^{-1/(\sigma\nu z)}, h/|r|^{\beta\delta}), \tag{3.13}$$

where $\alpha = 2$, ν is the correlation length exponent ($\xi \sim r^{-\nu}$), and z is the dynamic exponent relating the correlation length to the characteristic time ($\xi \sim T_0^z$).

Mean-field theory has the advantage of being easily tractable and provides a qualitative picture of the behavior of the model, but the numerical values of the exponents are typically inaccurate in dimensions lower than the upper critical dimension, which for the RFIM it is equal to $d_c = 6$. To overcome this problem, one can perform a renormalization group analysis with an expansion in $\epsilon = 6 - d$ as in equilibrium critical phenomena. The renormalization group is quite involved and we do not discuss it here in detail; we will only quote the main results (for a complete discussion the reader is referred to Dahmen and Sethna (1996)). The exponents determining the scaling of the order parameter have been computed to first order in ϵ and are estimated to be $\beta = 1/2 - \epsilon/6$ and $\delta = 3 + \epsilon$. The avalanche exponent τ displays only corrections to order ϵ^2 and the cutoff exponent is estimated as $\sigma = 1/2 - \epsilon/12$.

Large-scale numerical simulations have been performed to obtain reliable estimates for the critical exponents in various lattice dimensions and the results compare well with the RG predictions (see Fig. 3.7). The avalanche size distribution is most naturally computed by integrating the avalanches over the entire hysteresis loop and, with the exponent τ_{int} measured in this way, can be related to the other exponents by a scaling relation $\tau_{int} = \tau + \sigma\beta\delta$. A similar relation holds true for the duration exponent α_{int}. Typically, the distribution displays an initial power-law behavior and a cutoff that depends on the disorder R (see Fig. 3.8). The distribution for different values of R can then be collapsed using a scaling form of the type of Eq. 3.11 (Perkovic *et al.* 1995,1999). The best numerical estimate for the exponents (or exponent combinations) for $d = 3$ is reported in Table 3.1, together with the mean-field values.

The RFIM has been extensively simulated in $d = 2$, using extremely large system sizes (up to $(7000)^2$) (Perkovic *et al.*, 1995; Perkovic *et al.*, 1999). Despite this effort, it is not possible to obtain a reliable estimate of the threshold R_c for the disorder-induced transition. Notice that in $d = 1$ the transition is absent (i.e $R_c = 0$), as

	τ	α	$\sigma\nu z$	$1/\sigma$	τ_{int}	α_{int}	β	$1/\nu$
$d = 3$	1.60	2.05	0.57	4.2	2.03	2.81	0.0353	0.71
MF	3/2	2	1/2	2	2	3	1/2	2

Table 3.1 The numerical values of the exponents for the RFIM in $d = 3$ compared with the results of mean-field (MF) theory.

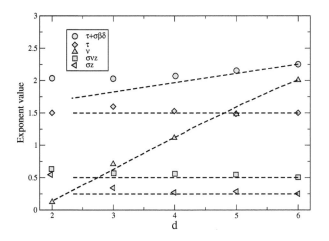

Fig. 3.7 Comparison between the RG predictions for the critical exponents of the RFIM and the results of numerical simulations in various dimensions. Source: Data obtained from Perkovic *et al.* 1999, Fig. 14.

shown by solving the model exactly (Shukla, 2000; Dante *et al.*, 2002). In $d = 2$ the simulations are consistent either with $R_c = 0$ or with very large correlated critical exponents (i.e. $\nu = 5.3 \pm 1.4$, $1/\sigma = 10 \pm 2$ with $R_c = 0.54$). An avalanche scaling exponent has also been estimated as $\tau_{int} = 2.04 \pm 0.04$. It is highly likely that $d = 2$ is indeed the lower critical dimension, as is expected for these models in equilibrium (Imry and Ma, 1975) and from the analysis of domain wall dynamics discussed in section 3.5. If this is the case, power-law scaling may still be observed for low disorder, with the conjecture that $\tau = 3/2$ (Perkovic *et al.*, 1995).

The RFIM represents probably the simplest model to study disorder induced transitions, but other forms of disorder are possible and the natural question is whether the transition is universal. Numerical simulations indicate a close similarity between the exponents, for the case of random bonds (Vives and Planes, 1994; Bertotti and Pasquale, 1990), site dilution (Vives and Planes, 2000), and random anisotropies (Vives and Planes, 2001). This agreement can be justified in the latter case using renormalization group arguments (da Silveira and Zapperi, 2004).

The situation is different, however, for spin glasses, where the coupling between the spins can be both ferromagnetic and antiferromagnetic. The prototypical spin glass model is due to Edwards and Anderson (1975) who introduced a random bond Ising model

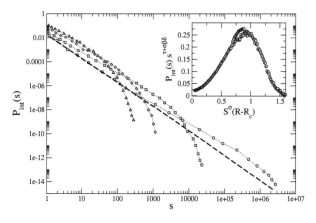

Fig. 3.8 The integrated avalanche distribution in the RFIM in $d = 3$ simulated in a lattice of 320^3 spins for different values of R. The inset shows the data collapse. Notice that the distributions are distorted by the scaling function and a power-law fit would not yield the asymptotic result. Data obtained from Perkovic *et al.* 1995, Fig. 1]

$$\mathcal{H} = -\sum_{\langle i,j \rangle} J_{ij} s_i s_j - \sum_i H s_i, \tag{3.14}$$

where J_{ij} is extracted from a Gaussian distribution with zero mean and the first sum is restricted to nearest-neighbor sites. The mean-field limit of the Edwards-Anderson model is known as the Sherrington-Kirkpatrik (SK) model (Sherrington and Kirkpatrick, 1975) and corresponds to Eq. 3.14, but the sum is extended to all possible pairs of spin. A study of the field-driven SK model shows a hysteresis loop with power-law-distributed avalanches limited only by the system size and independent of the disorder distribution (Pázmándi *et al.*, 1999). Similar results were obtained studying the static avalanches associated with the changes in the model ground states as a function of the applied field (Le Doussal *et al.*, 2010).

Finally, another interesting variant of the driven RFIM was introduced to model avalanche fluctuations in the resistance drops during a temperature-driven metal-insulator transition in polycrystalline films (Shekhawat *et al.*, 2011). The coupling in the model is long-ranged and anisotropic and produces a transition between a bolt-like phase, similar to dielectric breakdown, and a percolation phase. At the boundary between the two phases, one observes power-law-distributed avalanches (Shekhawat *et al.*, 2011) in good agreement with experiments (Sharoni *et al.*, 2008).

3.5 Domain wall motion: Roughness and percolation

The disordered-induced transition in the RFIM is due to the competition between domain growth, prevalent at low disorder, and the nucleation of new domains, dominating for strong disorder. The role of disorder can also be understood by analyzing the growth of single domain walls as the disorder strength is varied. Robbins and coworkers (Ji and Robbins, 1991; Ji and Robbins, 1992; Koiller *et al.*, 1992; Koiller

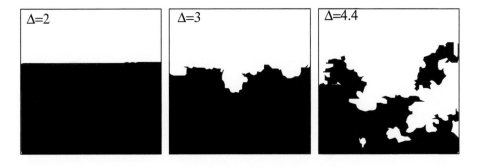

Fig. 3.9 The morphology of the invading front in the three-dimensional RFIM with uniform disorder as a function of Δ. The faceted growth phase is followed by a rough phase and finally by fractal phase. Source: Data obtained from Ji and Robbins (1992), Fig. 2.

and Robbins, 2000) have studied this problem in great detail using the dynamics encoded in Eq. 3.7 but only allowing for spin flips close to the domain wall, the interface between up and down spin. The simulation starts from a flat domain wall (a line in $d = 2$ and a plane in $d = 3$) where spins are flipped until the cluster spans the lattice from end to end. In most simulations, disorder is chosen in the form of random fields uniformly distributed in $[-\Delta, \Delta]$ (Ji and Robbins, 1992; Koiller *et al.*, 1992), Gaussian (Koiller and Robbins, 2000), or with additional random bonds (Ji and Robbins, 1991). By increasing the externally applied field, the domain wall is initially blocked until the field reaches the coercive value at H_c when the domain wall moves, and eventually spans the entire lattice. This is an example of *depinning transition* that will be studied in more detail in the next chapter.

In $d = 2$, increasing the disorder strength one observes a transition between a faceted growth, where the domain wall remains essentially flat with some steps, and fractal growth, where the front breaks up producing a self-similar cluster (Koiller *et al.*, 1992). It has been argued that this transition in $d = 2$ is only seen because of a lattice effect induced by the bounded disorder distribution (Drossel and Dahmen, 1998). Using a Gaussian distribution and a combination of numerical simulations and analytical results allows us to show that the transition at non-vanishing disorder is only a crossover effect. For sufficiently large lattices, fractal growth should prevail for any disorder values and faceted growth will only exist without disorder. Additionally, the scaling exponents, in particular the fractal dimension of the cluster of flipped spins, are conjectured to be equivalent to those of uncorrelated percolation. The general picture of the interface propagation problem closely resembles the general understanding of the hysteresis properties: There is no disorder-induced phase transition in $d = 2$ and disorder always dominates over the interactions.

The picture in $d = 3$ is more interesting, since, even for a bounded disorder distribution, we have a *self-affine* phase, where the geometrical scaling is anisotropic (as discussed in detail in the next chapter), and separates faceted fronts from fractal growth (see Fig. 3.9) (Ji and Robbins, 1992). For Gaussian disorders, the transition from self-affine to self-similar growth has been characterized in Koiller and Robbins (2000). The critical disorder separating the two phases was identified as $R_c = 2.5 \pm 0.2$

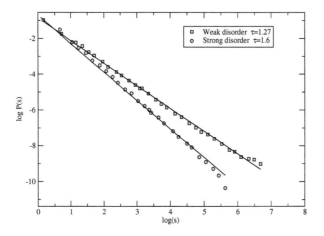

Fig. 3.10 The distribution of avalanche sizes for weak and strong disorder for the interface dynamics of the RFIM in $d = 3$. Source: Data obtained from Ji and Robbins 1992, FigS. 8 and 9.

and the correlation length exponent was found to be $\nu = 2.4 \pm 0.4$. Both the exponent and the critical point differs from the dynamics discussed in section 3.3. This difference is probably due to the role played by the spin flipping ahead of the interface in the model of Sethna *et al.* (1993).

In both the self-affine and the self-similar regimes, the interface moves by avalanches of flipped spins. At the depinning field H_c, the distribution of the avalanche volumes s decays as a power law

$$P(s) = s^{-\tau}, \tag{3.15}$$

where, in $d = 3$ $\tau = 1.28 \pm 0.05$ in the self-affine phase and $\tau = 1.59 \pm 0.05$ as shown in Fig. 3.10 (Ji and Robbins, 1992). The first value will be explained in the next chapter by the theory of interface depinning, while the second is remarkably close to the cluster distribution of percolation (see section 3.2). This was already noted by Ji and Robbins (1992), who pointed out the similarity between a set of exponents of the present problem with those of percolation.

4

The Depinning Transition

Several systems in condensed-matter physics can be described by elastic manifolds in random media. Concrete examples are provided by domain walls in ferromagnets, flux lines in type-II superconductors, contact lines, crack fronts, and dislocations. When an elastic manifold is pushed through a disordered landscape, it typically displays a depinning transition between a moving and a pinned phase. Over the past decades, a vast theoretical effort has been devoted to understand the depinning transition as a non-equilibrium-critical phenomenon. The morphology of the manifold is generally self-affine and can be characterized by a roughness exponent. Other scaling exponents have been introduced to characterize the behavior of correlation length, correlation time, and average velocity. In addition, the dynamics of elastic manifolds proceeds by avalanches that are power-law distributed at the depinning transition. Quantitative predictions of the critical exponents have been obtained analytically by the renormalization group and have been confirmed by numerical simulations.

4.1   Elastic manifolds in disordered media

Several phenomena in condensed-matter physics can be described as elastic manifolds moving in a disordered medium. Concrete examples are provided by domain walls in ferromagnets, flux lines in type II superconductors, contact lines, crack fronts, and dislocations. As we will illustrate in the present chapter, the interplay between disorder and elasticity leads to intriguing non-linear phenomena that are complex to treat theoretically. In general, an elastic manifold can be described by an N component displacement field $\mathbf{u} = (u_1(\mathbf{x}), \ldots u_N(\mathbf{x}))$ depending in general on d spatial coordinates $\mathbf{x} = (x_1, x_2, \ldots x_d)$. The displacement field measures the local deviation of the manifold from its equilibrium condition, which for simplicity we consider to be a flat manifold $u_i = 0$. A domain wall wall in the RFIM, discussed in section 3.5, is an example of a one-component ($N = 1$) elastic manifold in a d dimensional space, where $d = 1$ for the two-dimensional RFIM and $d = 2$ for the three-dimensional RFIM. In the RFIM, the internal dimension of the manifold d is smaller than the physical dimension $d' = d + 1$ in which the manifold is embedded. In this chapter, we will restrict our attention to single-valued displacement fields, so that our discussion will not apply to the strong disorder phase where domain walls have overhangs or branches.

Phase boundaries such as magnetic domain walls (Lemerle *et al.*, 1998; Zapperi *et al.*, 1998), but also grain boundaries or contact lines (Rolley *et al.*, 1998), are all defined by $N = 1$ and $d' = d + 1$. Similarly, elastic lines in two dimensions and membranes in three dimensions all have $N = 1$ (see Fig. 4.1). On the other hand,

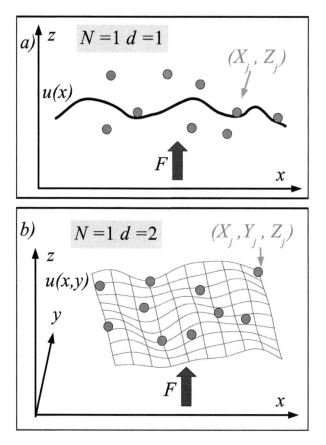

Fig. 4.1 Examples of elastic manifolds in random media. (a) A line in two dimensions corresponds to $d = 1$ and $N = 1$. A constant applied force F drives the manifold across a landscape of pinning centers placed at $(\mathbf{X}_j, Z_j)$. (b) A membrane embedded in three dimensions corresponds to the case $d = 2$ and $N = 1$.

individual superconducting vortices and crystal dislocations are lines moving in three dimensions and are thus characterized by $N = 2$ and $d' = d + 2$. Finally, three-dimensional elastic bodies are defined by $N = 3$, $d = 3$, and $d' = d = 3$. In this section, for the sake of simplicity, we will concentrate on the $N = 1$ case, but most of what we write could easily be generalized to $N > 1$.

Deforming an elastic manifold from its equilibrium position involves an energy cost that, for small deformations and short-range interactions, can be approximated as

$$E_{el} = \frac{C}{2} \int d^d x (\nabla u)^2, \tag{4.1}$$

where we have ignored higher-energy terms in the displacement field. Notice that the elastic energy is written in Eq. 4.1 as a function of the spatial derivatives of the displacement field u, ensuring that a rigid translation of the manifold (i.e. $u(\mathbf{x}) \rightarrow u(\mathbf{x}) + a$) would have no energy cost.

When regions along the manifolds interact through long-range forces—imagine for instance an electrically charged polymer chain—the elastic energy may become non-local and Eq. 4.1 should be replaced by

$$E_{lr} = K_0 \int d^d x d^d x' \frac{(u(x) - u(x'))^2}{|x - x'|^{d+\mu}},\qquad(4.2)$$

where $1 \leq \mu < 2$ is an exponent characterizing the long-range interactions.

The elastic energy in Eq. 4.1 can also be expressed in Fourier space as

$$E_{el} = \frac{C}{2} \int \frac{d^d q}{(2\pi)^d} |q|^2 |u_q|^2.\qquad(4.3)$$

while the long-range energy Eq. 4.2 can be written if Fourier space as

$$E_{lr} \propto K_0 \int \frac{d^d q}{(2\pi)^d} |q|^\mu |u_q|^2.\qquad(4.4)$$

The formal similarities between the two Fourier space expressions, suggests that for $\mu \rightarrow 2$, the long-range energy in Eq. 4.1 should approach the short-range energy reported in Eq. 4.4. A more rigorous calculation of the Fourier transform of Eq. 4.2 shows, however, that the long-range energy displays an additional logarithmic factor in Fourier space.

When the elastic manifold evolves in a disordered medium, we should consider an additional random potential $V(\mathbf{x}, z)$ describing the interactions between the manifold and the disorder. Different forms of disorder are conceivable, each corresponding to a specific form of the random potential. For instance, a collection of randomly placed defects (see Fig. 4.1) could be described by a potential given by the sum of the contributions associated with each defect

$$V(\mathbf{x}, u) = \sum_j V^{(j)}(\mathbf{x} - \mathbf{X}_j, u - Z_j),\qquad(4.5)$$

where $V^{(j)}$ measures the interaction between the manifold and the defect j located at position $(\mathbf{X}_j, Z_j)$. If the individual defects are all similar, we could use the same potential function for all of them (i.e. $V^{(j)}(\mathbf{x}, u) = V_0(\mathbf{x}, u)$). Depending on the system at hand, the interaction potential could be either attractive, so that each region of the manifold would prefer to sit on top of the nearest defect, or repulsive, so that the manifold would try to avoid the defects as much as possible.

At a coarse-grained scale where individual defects are not visible, the random potential could simply be described by a random function with zero mean and short-range correlations of the type

$$\langle V(\mathbf{x}, u) V(\mathbf{x}', u') \rangle = \delta^d(\mathbf{x} - \mathbf{x}') R((u - u')/\xi_p),\qquad(4.6)$$

where $R(u)$ is a rapidly decaying function and ξ_p is the typical disorder correlation length. In any case, the total energy $E_{tot} = E_{el} + \int d^d x V$ is a complicated non-linear function of the displacement field. The complications stem manly from the u

dependence in the random potential $V(\mathbf{x}, u)$, which makes the system hard to treat mathematically and leads to interesting non-trivial behavior. The manifold is subject to competing interactions: elasticity tends to flatten the manifold, while disorder tends to deform it. As a consequence of this competition, at equilibrium the manifold displays a rough shape, described by self-affine geometry.

While the equilibrium properties of elastic manifolds in disordered media are extremely interesting and non-trivial, we concentrate here on their non-equilibrium dynamics in response to an externally applied force. As we will discuss in more details in the coming chapters, an external force might be provided by magnetic field s in the case of domain walls in ferromagnets, by mechanical stresses for dislocation lines or crack fronts, and by electric fields for superconducting vortices. In all these examples, inertia can safely be neglected and we can therefore describe the moving manifold by an overdamped equation of motion

$$\Gamma \frac{\partial u(\mathbf{r}, t)}{\partial t} = -\frac{\delta E_{tot}(\{u(\mathbf{r}, t)\})}{\delta u(\mathbf{r}, t)} + F, \tag{4.7}$$

where F is a constant external force and Γ is a damping constant. Eq. 4.7 contains a functional derivative of the total energy of the manifold. In the case of short-range forces, Eq. 4.7 reduces to

$$\Gamma \frac{\partial u(\mathbf{r}, t)}{\partial t} = C\nabla^2 u(\mathbf{r}, t) + f(\mathbf{r}, u) + F, \tag{4.8}$$

where f is the pinning field obtained from the pinning potential V. The model in Eq. 4.8 is also known as the *linear interface model* (LIM), although the equation is not strictly linear because of the quenched disorder. Other forms of non-linearities can be considered, such as the Kardar–Parisi–Zhang (KPZ) term $f_{KPZ} = \lambda(\nabla u)^2$ that can be derived considering the lateral motion of the manifold (Kardar *et al.*, 1986):

$$\Gamma \frac{\partial u(\mathbf{r}, t)}{\partial t} = C\nabla^2 u(\mathbf{r}, t) + \lambda(\nabla u)^2 + f(\mathbf{r}, u) + F, \tag{4.9}$$

Eq. 4.9, known as the quenched KPZ equation (Q-KPZ), can not be derived from a Hamiltonian. A non-linear Hamiltonian model can be obtained if we include the next order in the gradient expansion of the elastic energy, namely $E_{nl} = A/4 \int d^d x (\nabla u)^4$. This leads to a non-linear elastic model (NLEM) described by the following equation of motion (Rosso and Krauth, 2001)

$$\Gamma \frac{\partial u(\mathbf{r}, t)}{\partial t} = C\nabla^2 u(\mathbf{r}, t) + A\nabla^2 u(\nabla u)^2 + f(\mathbf{r}, u) + F. \tag{4.10}$$

From numerical simulations, it appears that the two non-linear models, the Q-KPZ and the NLEM, fall into the same universality class, as we also discuss in what follows.

4.2   Interacting particles in random media

In the previous section, we defined the dynamics of elastic manifolds in random media. A very similar description applies to a collection of interacting particles moving in a disordered landscape, provided for example by a set of randomly placed immobile defects. For the sake of simplicity, we restrict our attention to a set of N_0 identical particles interacting by a pairwise potential. In this case, identifying the particle coordinates by $\mathbf{r}_i$, where $i = 1, ...N_0$, the equations of motion are given by

$$M\frac{d^2\mathbf{r}_i}{dt^2} + \Gamma\frac{d\mathbf{r}_i}{dt} = \sum_j \mathbf{F}_{int}(\mathbf{r}_j - \mathbf{r}_i) + \mathbf{F}_{ext} + \mathbf{F}_p(\mathbf{r}), \tag{4.11}$$

where M is the mass of the particles and Γ is a damping coefficient. In most cases of interest, dissipation is so strong that we can safely neglect inertia, setting $M = 0$ in Eq. 4.11.

The interaction force $\mathbf{F}_{int} = -\nabla V_{int}(\mathbf{r})$ is derived from the interparticle potential V_{int} that can have different forms, the simplest case being a short-range repulsive central force $\mathbf{F}_{int}(\mathbf{r}) = \hat{r}K(|\mathbf{r}|/\xi_{int})$. This force can be characterized by its peak value $K(0)$ and its range ξ_{int}. Increasing the complexity of the model, one can consider non-monotonic interactions (i.e. the force can be repulsive or attractive depending on the distance), long-range interactions (i.e. $K(r) \sim r^{-(d+\alpha)}$ with $0 < \alpha < 2$, for large r), anisotropic forces (i.e. $\mathbf{F}_{int} = \hat{r}K(\mathbf{r})$), non-central forces (i.e. $\mathbf{F}_{int}(\mathbf{r}) \times \hat{r} \neq 0$), and different combinations of the above. The second term in the right-hand side of Eq. 4.11 represents the external force, which we consider constant for simplicity, and the third term is due to the disorder. As an illustration, we can consider a set of N_p immobile pinning centers placed randomly at position $\mathbf{R}_j$, giving rise to a random force field

$$F_p(\mathbf{r}) = \sum_j \mathbf{f}_p((\mathbf{r} - \mathbf{R}_j)/\xi_p), \tag{4.12}$$

where ξ_p is the range of the individual pinning forces. Normally, the particular shape of the pinning potential does not matter as long as its range is short. Finally, thermal effects can be included by adding a random uncorrelated Gaussian noise $\eta(\mathbf{r}, t)$ to the right-hand side of Eq. 4.11.

Once the interactions between the particles and the other external forces have been specified, we should also define the boundary and initial conditions of the model. A common choice is to use periodic boundary conditions, and to place the particles in their zero-temperature equilibrium positions (i.e. a crystal). Alternatively, particles can be placed randomly in the system, mimicking a sudden quench of the system from a disordered high-temperature phase. These conditions are appropriate if we are interested in modeling the dynamics in the bulk of the material, without worrying about surface effects. On the other hand, in some cases boundary effects are at the core of the phenomena and one should then implement different initial and boundary conditions. Consider for instance the case of flux lines entering a type-II superconductor as an external magnetic field is increased.

Periodic boundary conditions need special care when interactions are long ranged, since in this case one cannot impose a cutoff to the extent of the interaction force, as is often done for short range-forces. One should instead consider explicitly the interaction of the particles in a given finite cell with all the periodic images of the system. The infinite sum over all the images can rarely be performed exactly, but a

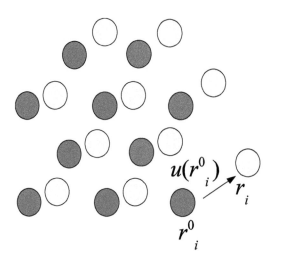

Fig. 4.2 The displacement of a set of interacting particle in two dimensions. In the absence of topological defects, there is a one-to-one correspondence between particles before and after deformation, and one can define a displacement vector with a well-defined continuum limit.

simple truncation of the series provides a rather poor approximation since the sum is slowly converging. To overcome this problem one can employ the Ewald summation method (Ewald, 1921), originally proposed for Coulomb interactions, after generalizing it for the appropriate interactions involved.

It is possible to study an interacting particle system in the *elastic* approximation assuming that the pinning forces are not strong enough to break the topological properties of system (Le Doussal and Giamarchi, 1998). If particles are arranged into a crystal, this assumption implies that the external force and the disorder preserve the topological order and no defects are generated. If no topological defects exist, there is a one-to-one relation between the positions of the particles before and after deformation (see Fig. 4.2). Hence, we can define displacement vectors $\mathbf{u}(\mathbf{r}_i^0) = \mathbf{r}_i - \mathbf{r}_i^0$, where $\mathbf{r}_i$ are the coordinates in the deformed system and $\mathbf{r}_i^0$ are the equilibrium positions. The particle interaction energy can then be expanded in terms of the displacement field, which is assumed to be small:

$$E_{el} \equiv \sum_{ij} V(\mathbf{r}_i - \mathbf{r_j}) \simeq U_0 + \sum_{\alpha,\beta} \sum_{ij} (u_\alpha(\mathbf{r}_i^0) - u_\alpha(\mathbf{r}_j^0)) \frac{\partial^2 V}{\partial r_\alpha r_\beta} (u_\beta(\mathbf{r}_i^0) - u_\beta(\mathbf{r}_j^0)), \quad (4.13)$$

where U_0 is the energy in equilibrium. We can then take the continuum limit expanding for small gradients $u(\mathbf{r}) \simeq u(\mathbf{r}^0) + (\mathbf{r} - \mathbf{r}^0) \cdot \nabla u(\mathbf{r})$ and obtain

$$E_{el} = E_0 + \frac{1}{2} \int d^3r \sum_{\alpha\beta\gamma\delta} \Lambda_{\alpha,\beta\gamma\delta} \frac{\partial u_\alpha}{\partial r_\gamma} \frac{\partial u_\beta}{\partial r_\delta}, \quad (4.14)$$

where we have used Greek indices for tensor and vector components to avoid confusion with particle labels, denoted by Latin indices. The elastic tensor $\Lambda_{\alpha\beta\gamma\delta}$ can be expressed in terms of the interparticle pair potential as

$$\Lambda_{\alpha\beta\gamma\delta} = -\frac{1}{2}\sum_{r^0} r^0_\gamma r^0_\delta \frac{\partial^2 V}{\partial r_\alpha r_\beta}. \tag{4.15}$$

Clearly this expansion is a good approximation as long as the sum in Eq. 4.15 converges. If interactions are long-range, so that the interaction potential is a slowly decaying power law (i.e. $V(r) \sim r^{-(d-1+\mu)}$ with $\mu < 2$), the elastic energy cannot be expanded in gradients and we have to consider a non-local interaction kernel.

Coming back to the local limit, a further simplification is obtained in Eq. 4.14 considering the symmetries of the equilibrium system. In the case of an isotropic system, the elastic tensor has only two independent components and the energy reduces to

$$E_{el} = E_0 + \frac{1}{2}\sum_{\alpha,\beta}\int d^3r K\delta_{\alpha\beta}\left(\frac{\partial u_\alpha}{\partial r_\alpha}\right)^2 + G\left(\frac{\partial u_\alpha}{\partial r_\beta}\right)^2, \tag{4.16}$$

where K and G are the bulk and shear moduli, respectively.

Using the elastic expression for the interparticle energy, we can rewrite the equation of motion for the particles as

$$\frac{\partial u_\alpha}{\partial t} = \mu\nabla^2 u_\alpha + (K+G)\frac{\partial}{\partial r_\alpha}(\nabla \cdot \mathbf{u}) + F + f_\alpha(r,u). \tag{4.17}$$

Notice the formal similarity with Eq. 4.8, the only difference being in the vectorial structure of Eq. 4.17. It is also possible to expand the equation of motions around a moving reference state. In that case we would obtain the same equation with an extra convective term $\mathbf{v} \cdot \nabla u$ on the left-hand side (Giamarchi and Le Doussal, 1996; Balents *et al.*, 1997). In the case of long-range interactions, the gradients are replaced by a non-local interaction kernel and the equation of motion becomes

$$\frac{\partial u_\alpha}{\partial t} = \sum_{\alpha,\beta}\int d^dr' K_{\alpha\beta}(r-r')(u_\beta(r') - u_\beta(r)) + F + f_\alpha(r,u), \tag{4.18}$$

with $K_{\alpha\beta}(r) \sim 1/r^{d+\mu}$ for large distances.

When pinning forces become stronger and/or pinning centers more diluted the topological order of the system typically breaks down. In this case, it is not possible to describe the deformation in terms of a displacement field and plastic deformation should be explicitly considered. Due to these difficulties, a complete theoretical understanding of plastic depinning is still not available and one should rely on numerical simulations.

4.3 The pinning force

4.3.1 Weak pinning: Collective pinning theory

Elastic deformations help the manifold overcome pinning centers, so that the pinning force of an isolated defect often does not provide a good estimate of the force needed to depin the entire manifold. To treat the collective effects of many pinning centers on a deforming manifold, we can employ a quite successful and influential scaling theory, derived independently in different contexts ranging from flux lines in superconductors (Larkin, 1970) to dislocations in crystals (Labusch, 1970) and to domain walls in ferromagnets (Hilzinger and Kronmüller, 1976). See also Giamarchi (2009) for an extensive discussion of this topic.

The central concept of collective pinning theory is the identification of a typical length scale L_c, known as the Larkin length (Larkin, 1970), defined as the length at which the typical displacements u of the manifold are of the order of the correlation length ξ_p of the pinning potential. This is also the scale at which pinning forces exactly balance the elastic forces. Local deformations of the manifold are ruled by the competition between elasticity, favoring a flat conformation, and pinning forces, which induce local deformations as the manifold is attracted towards the randomly arranged pinning centers. We can estimate these two contributions by simple scaling arguments, considering a portion of the manifold of linear size L undergoing deformations of order $u \sim \xi_p$. For short-range elasticity (Eq. 4.1), the elastic energy scales as $E_{el} \sim CL^{d-2}\xi_p$. The typical value of the pinning energy is due to the fluctuations of the random potential since its average is zero. Hence, considering Eq. 4.6, the pinning energy scales as $E_{pin} \sim (R(0))^{1/2}L^{d/2}\xi_p^{-1/2}$. At the Larkin length, elastic and pinning energies should have the same magnitude, yielding

$$L_c \sim \left(\frac{C^2\xi_p^2}{R(0)}\right)^{\frac{1}{4-d}}. \tag{4.19}$$

Notice that the collective pinning length decreases with the disorder strength as long as $d < d_c = 4$.

The dimension $d_c = 4$ enters here for the first time, but has in fact a great importance, being the upper critical dimension of the model: For $d > d_c$ the manifold does not roughen and the critical behavior at the depinning transition is described by mean-field theory. Since the physical dimension is usually lower than four, one may think that this limit is without practical interest. This is not true, because for long-range elasticity d_c can become equal to or even smaller than the physical dimension of the problem. To see this, we can repeat the Larkin argument for non-local elastic interactions with $\mu < 2$, for which the elastic energy scales as $E_{el} \sim K_0\xi_p^2L^{d-\mu}$. Hence, the collective pinning length becomes

$$L_c \sim \left(\frac{K_0^2\xi_p^2}{R(0)}\right)^{\frac{1}{2\mu-d}}, \tag{4.20}$$

and the upper critical dimension is now given by $d_c = 2\mu$.

The depinning force can then be estimated as the force needed to unpin a region of length L_c. This can be done by comparing the energy due to the external force, $E_{ext} = F\xi_pL^d$, to the pinning energy at the scale L_c, obtaining in the case of local elasticity

$$F_c \sim \frac{c\xi_p}{L_c^2}. \tag{4.21}$$

This is a remarkable results since it relates a static quantity such as the Larkin length to a dynamic quantity like the depinning force (Giamarchi, 2009).

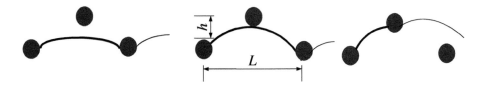

Fig. 4.3 The depinning of a line segment according to the strong pinning theory. Each segment bows between two pinning centers pushed by an external force. At depinning, a segment detaches from one pinning center and falls into the next one.

4.3.2 Strong pinning

Collective pinning theory holds as long as the pinning forces F_p are weak and the concentration of defects ρ_p is large, so that pinning is effectively ruled by the fluctuations of the defect concentration. Labusch (1969, 1970) recognized that this regime is valid as long as the dimensionless parameter $\kappa \equiv F_p/(4C\rho_p\xi_p^2)$ is much smaller than one. For large pinning forces and/or low concentrations, we enter into a different regime, often denoted as "strong pinning." In the strong pinning regime, the manifold can be pinned by individual defects and one can view a pinned line as a series of independent segments pinned at the edge by defects.

An estimate of the depinning force in the strong pinning regime was proposed by Friedel (1967), studying the bowing of a dislocation segment pinned between two solute atoms. The general idea is that depinning occurs when the most advanced part of the bowing segment encounters another pinning center. At the depinning point, Friedel assumed that each segment would encounter a new pinning center for each center left behind during the motion (Fig. 4.3). Based on this assumption, the area spanned by the line during this process should be equal to the inverse pinning point concentration: $Lu \simeq 1/\rho_p$ where L is the rest length of the dislocation segment and u is its maximum deformation. From simple geometrical considerations, we can then relate L and u to the curvature radius ρ (i.e. $2u\rho \simeq L^2$), which controls the elastic restoring force $F = C/\rho_p$ associated with the bowing. The depinning force is then obtained comparing the external force F with the pinning force F_p due to the solute atoms. Solving these four equations, we obtain the Friedel length

$$L_F \simeq (2C/\rho_p f_p)^{1/2}, \tag{4.22}$$

defined as the distance between defects on the line at depinning, and the depinning force

$$F_c \simeq (\rho_p f_p^3/2C). \tag{4.23}$$

4.4 Scaling and universality classes

4.4.1 Scaling exponents

The depinning of an elastic manifold moving in a disordered landscape is a very clear example of a non-equilibrium critical phenomenon. As in equilibrium phase transitions, the depinning transition is characterized by scaling laws and can be studied by the renormalization group method. In particular, just above the depinning transition the

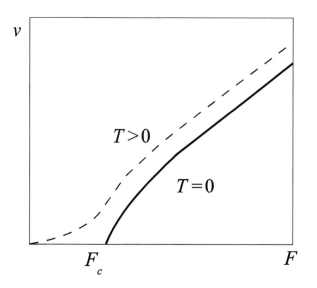

Fig. 4.4 A schematic view of the depinning transition. At $T = 0$ the velocity vanishes for $F < F_c$ and scales as a power law close to F_c. For $T > 0$ the depinning transition is *rounded* by the creep motion of the manifold.

average velocity of the manifold reaches a steady-state value v, playing the role of the order parameter of the transition and scaling with the distance from the critical point as

$$v \sim (F - F_c)^\beta, \tag{4.24}$$

where β is a critical exponent. This critical behavior is only valid at zero temperatures. Strictly speaking, at $T > 0$ the interface can depin due to termal activation, leading to creep motion (Giamarchi, 2009). In this case, one talks about a rounding of the phase transition, since the velocity is non zero for all the values of the force and the manifold is always moving for any positive force (see Fig. 4.4). At $T = 0$, depinning is associated with a diverging correlation length, in analogy with other second-order phase transitions in and out of equilibrium,

$$\xi \sim (F - F_c)^{-\nu}, \tag{4.25}$$

where ν is the correlation length exponent. The correlation length can be related to a correlation time t^* as

$$t^* \sim \xi^z, \tag{4.26}$$

where z is the dynamic exponent.

Scaling exponents also describe the morphology of the dislocation at the depinning transition: The typical displacement of the manifold scales with the system size as

$$\langle u^2 \rangle \sim L^{2\zeta}, \tag{4.27}$$

where L is the system size and ζ is the global roughness exponent. The same exponent can also be estimated from the averaged power spectrum $S(k)$ of the displacement function $u(x)$. The scaling for the power spectrum is expected to be $S(k) \sim k^{-(2\zeta+1)}$ Another common measure of the roughness exponent is provided by the displacement auto-correlation function

$$\langle (u(x) - u(0))^2 \rangle \sim |x|^{2\zeta_{loc}}, \tag{4.28}$$

where ζ_{loc} is the local roughness exponent. For self-affine manifolds, we expect $\zeta_{loc} = \zeta \leq 1$, but occasionally the two exponents can be different. In this case, we talk about anomalous scaling or super-roughening (Lopez, 1999). For instance, simulations of line depinning ($d = 1$, $N = 1$) with short-range elasticity yield $\zeta \simeq 1.2$ and $\zeta_{loc} = 1$ (Leschhorn, 1993; Leschhorn *et al.*, 1997). In most other relevant cases, namely $d > 1$ and $\mu < 2$ or for the non-linear model, one finds instead $\zeta_{loc} = \zeta$.

The exponents defined above are not all independent. A first scaling relation can be obtained by simple dimensional analysis, noticing that $v = du/dt$. Hence, we must have $\beta = \nu(\zeta - z)$. A second scaling relation, which is only valid in the linear model, derives from the fact that Eq. 4.8 is invariant under the transformation

$$u \rightarrow u + \nabla^{-2} g(x) \qquad F \rightarrow F - C g(x), \tag{4.29}$$

where ∇^{-2} is the inverse Laplacian and $g(x)$ is a generic function. Exploiting this symmetry, it is possible to derive the scaling relation

$$\nu = \frac{1}{2 - \zeta}. \tag{4.30}$$

For the long-range model with $\mu < 2$, this scaling relation is replaced by

$$\nu = \frac{1}{\mu - \zeta}. \tag{4.31}$$

In summary, we can describe the scaling behavior of the linear model using only two independent exponents, while for the non-linear model we need three.

4.4.2 Mean-field theory and the renormalization group

The mean-field theory provides a good qualitative description of the depinning transition (Fisher, 1985; Koplik and Levine, 1985; Leschhorn *et al.*, 1997). To derive mean-field equations (Narayan and Fisher, 1993), one studies an equation of motion of the LIM with an infinite ranged interaction kernel. To this end, it is convenient to first discretize Eq. 4.7

$$\frac{\partial u_i(t)}{\partial t} = F + \sum_j K_{ij}(u_j(t) - u_i(t)) + f_i(u), \tag{4.32}$$

where K_{ij} in Fourier space has the form

$$K(q) \propto q^{\mu}, \tag{4.33}$$

where $\mu = 2$ corresponds to the local model in Eq. 4.8. The infinite range model is obtained setting K_{ij} to a constant $\bar{K}$

$$\frac{\partial u_i(t)}{\partial t} = F + \bar{K}(\bar{u} - u_i(t)) + f_i(u), \tag{4.34}$$

where $\bar{h} \equiv \sum_i h_i/N$ is the center-of-mass displacement and N is the system size. Eq. 4.34 can be solved self-consistently to obtain the scaling behavior. The mean-field behavior depends on the shape of the random potential: For potentials containing cusps, one obtains that the velocity of the interface grows linearly for $F > F_c$, so that $\beta = 1$. If one uses instead a smooth potential, one obtains $\beta = 3/2$ (Fisher, 1985). This result is a peculiarity of mean-field theory, since in finite dimensions the scaling exponents do not depend on the shape of the random potential (Narayan and Fisher, 1993).

To go beyond mean-field theory, Narayan and Fisher (1993) and Nattermann *et al.* (1992) have devised a functional renormalization group scheme that allows one to obtain an estimate of the critical exponents and compute scaling functions using an $\epsilon = d_c - d$ expansion. The method is based on an expansion around mean-field theory, using the Martin–Siggia–Rose formalism (Martin *et al.*, 1973). Notice that in Narayan and Fisher (1993), it was argued that the correct expansion should be done around the mean-field model with a cusped potential. This point was later clarified by Chauve *et al.* (2000) that were able to prove that the cusp arises naturally under the renormalization group flow. The first step is to construct a generating functional for the response and correlation functions, introducing an auxiliary field $\hat{u}(x, t)$,

$$Z = \int (du)(d\hat{u}) \exp\left\{ i \int d^d x \, dt \hat{u} F(u, f) \right\}, \tag{4.35}$$

where

$$F(u, f) = \frac{\partial u(x, t)}{\partial t} - C\nabla^2 h(x, t) -$$

$$\int d^d x' \, K(x - x')(u(x', t) - u(x, t)) - f(x, h) - F, \tag{4.36}$$

where we have considered both the long-range and short-range interaction term Following Narayan and Fisher (1993), we introduce a new field:

$$\phi_i = \sum_j K_{ij} u_j \tag{4.37}$$

which represents a coarse-grained version of u, and a corresponding auxiliary field $\hat{\phi}$. After averaging over the disorder one obtains an effective generating functional

$$\bar{Z} = \int (d\phi)(d\hat{\phi}) \exp(\tilde{S}(\phi, \hat{\phi})), \tag{4.38}$$

whose saddle-point value corresponds with the mean-field theory. The effective action is given by

$$\tilde{S} = \int d^d x \, dt F \hat{\phi}(x, t)$$

$$+ \int \frac{d^d q \, d\omega}{(2\pi)^d} \hat{\phi}(-q, -\omega)(-i\omega + K_0 q^\mu + C q^2)\phi(q, \omega)$$

$$-\frac{1}{2} \int d^d x \, dt dt' \hat{\phi}(x, t) \Delta(vt - vt' + \phi(x, t) - \phi(x, t'))\hat{\phi}(x, t), \qquad (4.39)$$

where the function $\Delta(x)$ is the disorder correlation function. Narayan and Fisher carried out an expansion around the saddle point to obtain the correction to the mean-field theory. The technical difficulty lies in the fact that $\Delta(x)$ contains an infinite series of non-linear terms that are all relevant. Hence, a functional renormalization group is necessary, where the full disorder correlator is renormalized. We do not attempt to discuss in detail these technical points and we direct the interested reader to the relevant literature (Narayan and Fisher 1993; Leschhorn *et al.* 1997; Chauve *et al.* 2000,2001,Le Doussal *et al.* 2002).

Here, we just show how to find the upper critical dimension from simple power counting. If we rescale space by a factor b ($x = bx'$) to ensure that the Gaussian part of the effective action remains invariant, time and fields have to be rescaled as $t = b^z t'$, $\phi = b^\zeta \phi'$, $\hat{\phi} = b^{\theta - d}\hat{\phi}'$, and $F = b^{-1/\nu}F'$. We notice here that for $\mu < 2$ the Laplacian term is subdominant at large length scales and we can thus omit it from the power counting, yielding

$$z = \mu \quad \zeta = \frac{2\mu - d}{2} \quad \theta = \frac{d - 2\mu}{2} \quad \nu = \frac{2\mu}{d}. \qquad (4.40)$$

For $d > 2\mu$ all the non-linear terms decay to zero at large length scales and the effective theory becomes Gaussian, while for $d < 2\mu$ an infinite set of non-linear terms becomes relevant. The upper critical dimension is therefore $d_c = 2\mu$, as anticipated from the scaling argument discussed in section 4.3.1. The "bare" critical exponents reported above are only valid at the upper critical dimension, where they coincide with mean-field values. Below d_c, the renormalization group yields prediction for the critical exponents

$$\zeta = \frac{\epsilon}{3}, \quad z = \mu - \frac{2\epsilon}{9}, \quad \beta = 1 - \frac{\epsilon}{9}, \quad \nu = \frac{1}{\mu - \zeta} \qquad (4.41)$$

to first order in the $\epsilon = 2\mu - d$ expansion (Narayan and Fisher, 1993; Leschhorn *et al.*, 1997). These exponents were later computed to order ϵ^2 (Chauve *et al.*, 2001) and the results are reported in Table 4.1 for the short-range interaction case ($\mu = 2$).

4.4.3 Numerical simulations

The functional renormalization group provides a useful framework to characterize the depinning transition and estimate the numerical value of the critical exponents. The results have been compared with numerical simulations performed according to various methodologies, from the direct numerical integration of the equation of motion, for both elastic manifolds and interacting particle models, to the simulation of discrete

	d	ϵ	ϵ^2	Padé approximation	simulation
	3	0.33	0.38	0.38±0.02	0.34±0.01
ζ	2	0.67	0.86	0.82±0.1	0.75±0.02
	1	1.00	1.43	1.2±0.2	1.25±0.05
	3	0.89	0.85	0.84±0.01	0.84±0.02
β	2	0.78	0.62	0.53±0.15	0.64±0.02
	1	0.67	0.31	0.2±0.2	≈ 0.3
	3	0.58	0.61	0.62±0.01	0.6
ν	2	0.67	0.77	0.85±0.1	0.8
	1	0.75	0.98	1.25±0.3	1.33

Table 4.1 Exponents for the LIM with short-range forces obtained from the renormalization group to various orders in the ϵ expansion and through a Padé resummation of the expansion. Results of numerical simulations from are reported for comparison. Source: Chauve *et al.* (2001), Leschhorn *et al.* (1997)

models which allow us to explore a broader range of time and lengths scales. Thanks to this extensive numerical effort, the scaling exponents of the depinning transition are known with very good precision. Furthermore, simulations provided essential information on the non-linear model that is still unsolved by the renormalization group method.

A widely used strategy to simulate the depinning transition is based on discrete cellular automaton models, in which the manifold displacement u, the time t, and the space $\mathbf{x}$ are all discretized taking integer values (Leschhorn, 1993, 1996). For a given configuration $\{u_i\}$ of the manifold, we can compute the local force F_i at each site i by a discretized version of Eq. 4.8

$$F_i = F + \frac{C}{a^2} \sum_{nn} (u_{i+nn} - u_i) + f_i(u_i), \tag{4.42}$$

where the sum runs over all the nearest neighbors nn of the site i, a is the discretization scale, and $f_i(u_i)$ is a random force. The automaton dynamics stipulates that a site is either stable ($F_i < 0$) or unstable ($F_i > 0$). Unstable sites are advanced by one step $u_i = u_i + a$ and a new value is extracted pinning force $f_i(u_i)$. Thus in the cellular automaton model, the local velocity can only be equal to zero or one and no backward motion is allowed. This limitation is believed to be irrelevant for the universal properties of the model. The reason is that close to the depinning threshold the velocity is very small and only a small part of the manifold moves at each given time. The validity of this assumption has been tested by comparing with a direct integration of the continuum equations (Leschhorn, 1993, 1996, Cule and Hwa 1998; Lacombe *et al.* 2000). Furthermore, Rosso and Krauth have developed a more sophisticated algorithm that instead of simulating the dynamics searches for all the possible pinned configurations of the manifold in a finite system (Rosso and Krauth, 2001). The procedure can be repeated for different values of the force and of the system sizes, allowing for the study of finite-size scaling. The method is particularly effective to estimate the roughness exponent at the transition (Rosso *et al.*, 2003).

	d	DPD	Q-KPZ	NLEM
	3	0.38	-	0.25
ζ	2	0.48	-	0.45
	1	0.63	0.63	0.63

Table 4.2 Roughness exponents from numerical simulations of the DPD model (Amaral *et al.*, 1995), direct integration of the quenched KPZ model (available only in $d = 1$ (Leschhorn, 1996)), and of the non-linear elastic model (NLEM) (Rosso *et al.*, 2003) in various dimensions

The cellular automaton model can also be generalized to take into account long-range interactions or non-linear terms. The long-range model has been simulated in Tanguy *et al.* (1998) for different values of μ, showing that the results are in good agreement with the renormalization group predictions for a wide ranges of values (see Fig. 4.5). The special case $\mu = 1$ has been investigated in several other papers (Schmittbuhl *et al.*, 1995; Ramanathan and Fisher, 1998; Rosso and Krauth, 2002), due to the relevance of this case for a variety of systems, from cracks (Schmittbuhl *et al.*, 1995) to contact lines (Ertas and Kardar, 1994). The roughness exponent for $\mu = 1$ is found to be $\zeta = 0.35$ (Schmittbuhl *et al.*, 1995; Ramanathan and Fisher, 1998) for various cellular automata and $\zeta = 0.38$ from the algorithm developed in Rosso and Krauth (2002).

In the case of non-linear elasticity, the universality class is again completely different, as shown by numerical simulations (Rosso and Krauth, 2001). In $d = 1$ one finds that the roughness exponent of the model both with the KPZ term (Q-KPZ) (Leschhorn, 1996) and with the quartic elastic energy (NLEM) (Rosso and Krauth, 2001; Rosso *et al.*, 2003) is the same and equal to $\zeta = 0.63$. A similar value is found in a wide variety of lattice growth models including some form of Laplacian relaxation and other mechanisms implementing lateral motion (Buldyrev *et al.*, 1992; Tang and Leschhorn, 1992; Sneppen, 1992; Amaral *et al.*, 1995). A crucial point is that overhangs should be forbidden, either by a solid-on-solid rule (Tang and Leschhorn, 1992; Sneppen, 1992) or by another relaxation rule (Buldyrev *et al.*, 1992; Amaral *et al.*, 1995). The result is that the interface is single valued and one can argue that it corresponds to a directed percolation path. For these reasons, this universality class has been defined as *directed percolation depinning* or DPD (Buldyrev *et al.*, 1992; Tang and Leschhorn, 1992; Amaral *et al.*, 1995). In Table 4.2 we compare the roughness exponent ζ for the DPD model (Amaral *et al.*, 1995) with the result obtained by the numerical integration of the Q-KPZ equation (Leschhorn, 1996) and the simulations of the non-linear elastic model (Rosso *et al.*, 2003). From the numerical values of the exponents, it is likely that the all these non-linear models belong to the same universality class, although slight deviations are seen for high dimensions. Finally, it is interesting to remark that the DPD universality class should only be observable at large enough scales when the non-linear coupling becomes relevant. At small scales one could ignore the non-linear term and recover the scaling of the linear model. There is in fact a universal crossover function interpolating between linear and non-linear depinning models that can be quantified accurately from simulations (Chen *et al.*, 2015).

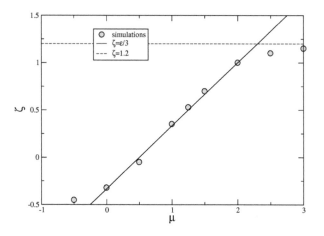

Fig. 4.5 The exponent ζ for different values of μ, obtained from numerical simulations of a model with a long-range interaction kernel, is compared with the prediction of the renormalization group at first order in the ϵ-expansion. For large μ the value approaches the value $\zeta = 1.2$ obtained for a short-range kernel. Source: Data from Tanguy *et al.* 1998.

4.5   Avalanche dynamics

4.5.1   Scaling relations

When the external force is increased monotonically and adiabatically, the manifold moves in avalanches of increasing size, up to the depinning point where the motion becomes continuous. We can derive scaling relations between the exponents characterizing the avalanche distributions and the other exponents discussed in section 4.4. Here, for simplicity we consider first the statistics of avalanches occurring at a constant value of $F < F_c$. The avalanche size distribution close to the depinning transition scales as

$$P(s) \sim s^{-\tau} f(s/s_0), \tag{4.43}$$

where $f(x)$ is a scaling function and the cutoff scales as $s_0 \sim (F - F_c)^{-1/\sigma}$ and is related to the correlation length ξ by

$$s_0 \sim \xi^{d+\zeta}, \tag{4.44}$$

where ζ is the roughness exponent (Fig. 4.6). Since the correlation length diverges at the depinning transition as in Eq. 4.25, we can derive a scaling relation

$$\frac{1}{\sigma} = \nu(d + \zeta). \tag{4.45}$$

The average avalanche size also diverges at the transition

$$\langle s \rangle \sim (F - F_c)^{-\gamma}, \tag{4.46}$$

where γ is related to τ and σ by

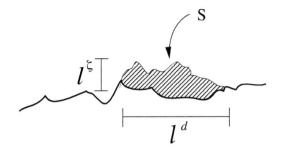

Fig. 4.6 A line moving between two pinned configurations in an avalanche of size $S \sim l^{d+\zeta}$

$$\gamma = \frac{(2-\tau)}{\sigma}. \tag{4.47}$$

An additional scaling relation can be obtained considering the susceptibility of the manifold to a variation of the external force, which is proportional to the average avalanche size (Narayan and Fisher, 1993). From power counting, it is easy to see that the susceptibility should scale as

$$\frac{d\langle u \rangle}{dF} \sim (F - F_c)^{-(1+\nu\zeta)}. \tag{4.48}$$

This relation together with Eq. 4.47 implies that

$$\tau = 2 - \frac{1+\nu\zeta}{\nu(d+\zeta)}. \tag{4.49}$$

In the case of the linear model, we can insert in Eq. 4.49 the scaling relation reported in Eq. 4.31, obtaining

$$\tau = 2 - \frac{\mu}{d+\zeta}. \tag{4.50}$$

One can also derive similar relations for the distribution of avalanche durations

$$P(T) \sim T^{-\alpha} g(T/T_0), \tag{4.51}$$

where the cutoff diverges at the transition as $T_0 \sim (F - F_c)^{-1/\Delta}$. From Eq. (4.25) and the relation $T_0 \sim \xi^z$, we obtain $\Delta = 1/z\nu$ and

$$\alpha = 1 + \frac{\nu d - 1}{z\nu}. \tag{4.52}$$

We note that all the relations (Eqs. 4.44–4.52) are valid as long as $d \leq d_c$. For $d > d_c$, one expects mean-field scaling to apply, for which $\tau = 3/2$, $\alpha = 2$, $\sigma = 2$, $\Delta = 1$. The values of the exponents τ and α for various universality classes, estimated using the cellular automaton model, are reported in Table 4.3.

We have derived scaling relations between avalanche critical exponents using simple arguments. These relations—Eq. 4.50 in particular—can be demonstrated more rigorously in the context of the functional renormalization group (Le Doussal *et al.*, 2009). Pierre Le Doussal, Kay Wiese, and coworkers have performed sophisticated calculations to compute explicitly several statistical properties of avalanches at the depinning transition (Le Doussal *et al.*, 2009; Rosso *et al.*, 2009; Le Doussal and Wiese, 2012; Le Doussal and Wiese, 2012; Le Doussal and Wiese, 2013; Dobrinevski *et al.*, 2015). In particular, they showed that the distribution of avalanche size can be written as

$$P(s) = \frac{\langle s \rangle}{s_m} p(s/s_m), \tag{4.53}$$

where $s_m = \langle s^2 \rangle / 2 \langle s \rangle$ and $p(x) = x^{-\tau} f(x)$ is a universal scaling function that can be computed using the ϵ expansion, yielding results in agreement with simulations (Rosso *et al.*, 2009). The scaling function describing the cutoff to the power-law behavior is computed to first order in ϵ

$$f(x) = \frac{A}{2\pi} \exp(C\sqrt{x} - B/4x^\delta), \tag{4.54}$$

where A, B, C and δ are all universal parameters that can be estimated by the renormalization group (Le Doussal *et al.*, 2009). Analytic predictions could also be obtained for other universal observables such as the duration and velocity distributions (Le Doussal and Wiese, 2013) and the average pulse shape (Dobrinevski *et al.*, 2015). Finally, it is worth recalling that, as already discussed in section 2.4, all these quantitative predictions for exponents and scaling functions can be applied directly to stochastic sandpiles thanks to an exact mapping between the field theories of the two problems (Le Doussal and Wiese, 2015).

		d	short-range	$\mu = 1$
		1	1.03	1.25
τ		2	1.27	1.5
		MF	3/2	3/2
		1	1.0	1.43
α		2	1.5	2
		MF	2	2

Table 4.3 The exponents for the decay of the distributions of avalanche sizes (τ) and durations (α) for the linear interface model (LIM) with short-range coupling and long-range interactions with $\mu = 1$. Source: Bonamy *et al.* (2008).

4.5.2 Driving modes

Avalanches normally do not occur under a constant applied force as discussed in the previous section, although this case could be experimentally relevant for low-temperature creep for F just below F_c. In many cases, the external force is ramped up at a slow constant rate up to the depinning point. In this case, the constant-force avalanche size distribution should be integrated over all the values of the forces

$$P_{int}(s) = \int_0^{F_c} dF P(s, F).$$ (4.55)

Using the scaling form in Eq. 4.43 and changing variables in the integral ($x = s^{\sigma}(F_c - F)$), we obtain

$$P_{int}(s) = \int_0^{F_c} dF s^{-\tau} f(s(F_c - F)^{1/\sigma}) = s^{-(\tau+\sigma)} \int_0^{F_c-F} dx f(x^{1/\sigma}).$$ (4.56)

Hence, the integrated size distribution scales with an exponent larger than τ, namely $\tau + \sigma$. Similarly, the integrated duration distribution scales with an exponent $\alpha + \Delta$

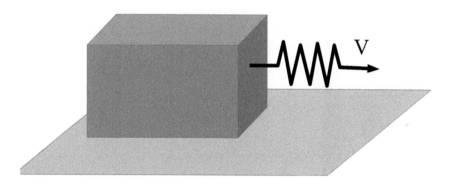

Fig. 4.7 A constant-velocity driving can be obtained by attaching the system to an elastic spring that is pulled at constant velocity, as in the friction setup schematized here.

In many experimentally relevant cases, researchers might want to control the velocity of the manifold rather than apply a constant force or a constant force rate. This is typically done by a suitable machine that enforces a constant velocity through an effective elastic spring. The classical example is provided by a friction force measuring apparatus in which a block is displaced over a substrate by attaching it to a spring pulled at constant velocity V (see Fig. 4.7). In the framework of the depinning transition, this driving mode corresponds to replace the force in Eq. 4.8 by

$$F = k(Vt - u(\mathbf{x}, t)),$$ (4.57)

where k is the stiffness of the spring. In this particular case, we assume that each point of the manifold is experiencing an effective spring force proportional to its displacement. Other possibilities can arise depending on the system. For instance, the spring

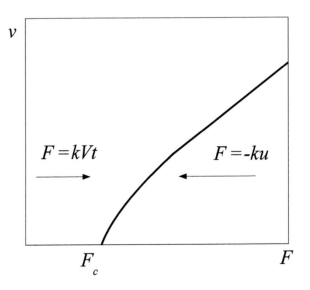

Fig. 4.8 A schematic view of the effect of constant veocity driving on the depinning transition. Below the depinning transition the manifold is pinned and the effective force increases, above the depinning transition the manifold moves and the effective force decreases.

may be attached to the edge of the manifold or to its center of mass which yields a force

$$F = k(Vt - \int d^d x u(\mathbf{x}, t)/L^d). \tag{4.58}$$

The general effect of this type of driving mode does not depend on the way the effective spring is attached, although there are some quantitative differences between the various cases as we will discuss below.

If we consider the general phase diagram of the depinning transition, we can readily understand how the external control of the velocity affects the dynamics of the manifold. If we start with $F = 0$ and then start to pull one end of the spring at a slow rate (i.e. $V \to 0$), the force on the manifolds slowly increases until it reaches the depinning force F_c. At this point, the manifold acquires a finite steady-state velocity and the displacement u grows. This leads to a decrease of the effective force which will eventually fall below the depinning force. Hence, the system is kept dynamically around the critical point (see Fig. 4.8). What we have just described is a basic mechanism acting in self-organized critical systems. As a result of this peculiar driving, the steady-state velocity signal is composed by avalanches that in the limit $V \to 0$ and $k \to 0$ are power-law distributed. A high driving velocity V would lead to overlapping avalanches, breaking up the scaling behavior. Similarly, the presence of a large spring constant k will hinder avalanche propagation. Examples of avalanche sequences are reported in Fig. 4.9 as obtained by numerical simulations of the cellular automata version of the LIM and the NLEM. Notice the large slopes observed in the LIM, signifying that $\zeta > 1$.

Fig. 4.9 A series of avalanches from simulations of a line moving in a disorered landscape. On the left panel is the linear interface model, showing a superrough interface with large slopes. On the right panel is the non-linear elastic model, displaying a self-affine structure. The grayscale indicates the time at which the avalanches occur, starting from the early avalanches at the bottom (in white) to the late avalanches at the top (in black).

The precise effect of the spring constant on the avalanche distribution cutoff s_0 can be described by a simple argument. Due to the restoring force provided by the spring, the avalanche propagation can not extend beyond a length scale ξ_k. In the case of the linear model, the avalanche cutoff can be obtained by comparing the typical strength of the interaction kernel over a length scale ξ_k, $f_{int} \sim K_0 u \xi_k^{-\mu}$, with the strength of the corresponding restoring force f_k. When the spring is attached to every point of the manifold as in Eq. 4.57, the restoring force scales as $f_k \sim ku$, yielding

$$\xi_k \sim \left(\frac{k}{K_0} \right)^{-\nu_k} \quad , \quad \nu_k = \frac{1}{\mu}. \tag{4.59}$$

In the case of a spring coupled to the center of mass of the manifold as in Eq. 4.58, the restoring force scales as $f_k \sim ku(\xi/L)^d$ yielding

$$\xi_k \sim \left(\frac{k}{K_0 L} \right)^{-\nu_k} \quad , \quad \nu_k = \frac{1}{\mu + d}. \tag{4.60}$$

The avalanche size and duration distributions cutoff s_0 and T_0 can then be obtained using the scaling relations reported above:

$$s_0 \sim \left(\frac{k}{K_0 L} \right)^{-1/\sigma_k} \quad , \quad 1/\sigma_k = \nu_k(d + \zeta) \tag{4.61}$$

and

$$T_0 \sim \left(\frac{k}{K_0 L}\right)^{-\Delta_k}, \qquad \Delta_k = z\nu_{k}, . \tag{4.62}$$

These scaling relations are confirmed by numerical simulations with the automaton model as shown in the data collapse of Fig. 4.10. For intance in the case of the LIM, we expect $\nu_k = 1/3$ and $1/\sigma_k \simeq 0.75$, as confirmed by simulations.

The case of boundary drive in which the force is only applied to one edge of the manifold needs a separate discussion (Paczuski and Boettcher, 1996). Consider for instance a linear elastic line that is pulled from one end and pinned at the other end, as illustrated in Fig. 4.11. In this case, we expect that the line itself will act as an elastic spring that mediates the dynamics. Hence, the same mechanism of self-organization described above is expected to apply and we expect the dynamics to be critical. The fact that the force is applied only at one end implies an additional scaling relation for the avalanche exponents. On average a small increase in the applied force F leads to an increase in the displacement of the order L. In other words, the susceptibility scales as $du/dF = \langle s \rangle = L$, which implies $\gamma/\nu = 1$. Inserting this result in Eq. 4.47, we obtain

$$\tau = 2 - \frac{1}{d + \zeta}, \tag{4.63}$$

which yields $\tau = 1.55$ in $d = 1$ and $\tau = 1.63$ in $d = 2$.

4.5.3 Extremal dynamics

We have illustrated possible mechanisms to drive a manifold across a disordered landscape, describing their effects on avalanche propagation. An additional driving method that has been studied extensively in the literature is the *extremal dynamics* (Sneppen, 1992; Zaitsev, 1992; Bak and Sneppen, 1993; Amaral et al., 1995; Paczuski et al., 1996). The idea is easily illustrated in the framework of the cellular automaton model described in section 4.4.3. Extremal dynamics implies that at each time step we chose the site with the highest local force and advance it by one step, a processes analogous to invasion percolation. With this prescription the average velocity of the manifold is constant and equal to $1/N$, where N is the number of sites. In the limit of $N \to \infty$, the manifold moves at vanishing velocity, approaching the depinning point from above. This can be seen by studying the distribution of local forces along the manifold. At the beginning of the simulation, starting from a flat manifold, the local force in each point of the manifold is given by the local pinning forces. As the manifold advances, however, weakly pinned parts of the manifolds are unpinned. Hence the local force distribution develops a gap. This process is repeated until the local force distribution becomes stationary displaying a gap in correspondence to the threshold force F_c (Sneppen, 1992; Zaitsev, 1992; Bak and Sneppen, 1993; Amaral et al., 1995; Paczuski et al., 1996).

Under extremal dynamics, avalanches can be defined considering the time evolution of the maximum local force f_k^{max} at step k. The avalanche size is defined as the number of subsequent moves for which $f_{k+1}^{max} < f_k^{max}$. In order to define a suitable time variable under extremal dynamics, we can associate to each updating step a time interval $\Delta t = 1/n_{act}$, where n_{act} is the number of "active" sites for which $f_i < F_c$

(Paczuski *et al.*, 1996). Using this definition, one can compute a set of scaling exponents and describe the spatio-temporal dynamics of the model (Paczuski *et al.*, 1996). The exponents defined at the critical point, such as τ or ζ, can be mapped directly into the corresponding ones measured at the depinning transition. Other exponents may instead depend on the peculiarities of the extremal dynamics and do not have a direct analog in the depinning case.

The interest in extremal models stemmed mostly from their peculiar mathematical properties, leading to a self-organization to the depinning point, rather than their applications to realistic physical examples. Avalanching systems are typically not extremal, since after the avalanche is started, possibly by an extremal process, it then propagates collectively and in parallel. Yet, some form of extremal dynamics can be realized in low-temperature creep (Zaitsev, 1992). One can imagine a Monte Carlo model for a manifold where the site i would advance at a rate

$$w_i \propto e^{-\frac{\Delta E_i}{k_B T}}, \tag{4.64}$$

where $\Delta E_i = -af_i$ is the energy difference between the states before and after the move and a is the discretization step. In the limit $T \to 0^+$, only the site with a maximum force f_i will have a non-vanishing rate to move and the dynamics would be equivalent to an extremal model. The only problem with this mapping is that it does not consider backward motion, which is necessary to ensure detailed balance. One could introduce a small positive force F biasing the motion and then consider the limits $T \to 0^+$ and then $F \to 0^+$.

(a)

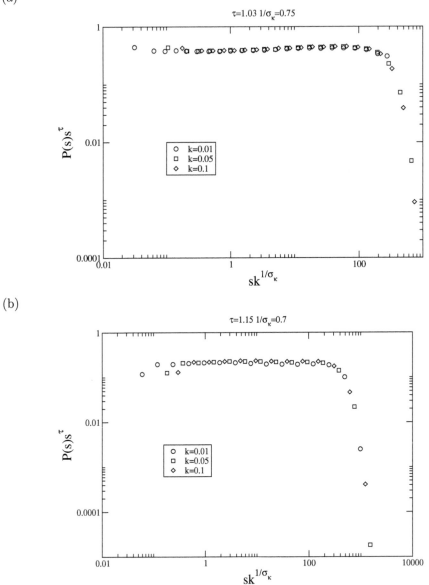

Fig. 4.10 Data collapse for the avalanche distributions of the linear interface model (a) and the non-linear interface model (b) in $d = 1$. The manifold is driven by a spring attached to its center of mass. The exponents are $\tau = 1.03$ and $1/\sigma_k = 0.75$ for the LIM and $\tau = 1.18$ and $1/\sigma_k = 0.7$ for the NLEM.

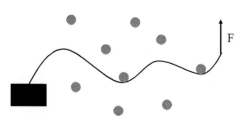

Fig. 4.11 An elastic line pulled through a disordered landscape by applying the force on one end while the other end is kept fixed.

5

Fracture

Understanding how materials break is still a fundamental problem of science and engineering. The statistical properties of fracture have attracted a wide interest across the statistical physics community. In this context, fracture is considered an irreversible process ruled by long-range interactions and disorder. Several experiments have revealed that fracture is indeed a complex phenomenon, described by scale-invariant laws. Notable examples include the acoustic emission activity prior to fracture, which typically displays an intermittent character and a power-law amplitude distribution.. To address this problem, we start from the theory of elasticity and consider the elastic stresses associated with a crack propagating through a disordered medium. Whenever damage is diffusive, however, studying a single crack is not adequate and one usually resorts to lattice models, from simple mean-fields like fiber bundles to more complicated and realistic models of disordered elastic media. We conclude by making direct analogies between fracture and phase transitions.

5.1   Elements of linear elastic theory

Small deformations of a solid are usually described in terms of the theory of linear elasticity and a crack can be seen as a boundary condition to the elastic equations. In order to set the basic notation, we recall here a few elements of the theory of elasticity. Given the local displacement field, $\mathbf{u}(\mathbf{r})$, it is useful to define the symmetric *strain tensor* as

$$\epsilon_{ik} \equiv \frac{1}{2} \left(\frac{\partial u_i}{\partial x_k} + \frac{\partial u_k}{\partial x_i} \right). \tag{5.1}$$

When a solid deforms, interactions at the molecular scale provide restoring forces that tend to bring the solid back to static equilibrium. Due to the action-reaction principle, the resulting force acting on the volume element V, $\int_V dV \mathbf{F}$ is only given by surface terms. Thus each component of $\mathbf{F}$ is the divergence of a vector field (i.e $F_i = \sum_k \frac{\partial \sigma_{ik}}{\partial x_k}$) so that

$$\int_V dV F_i = \int_S dS \sum_k n_k \sigma_{ik}, \tag{5.2}$$

where S is the surface of the volume V, n_k is a unit vector normal to S and σ_{ik} is the *stress tensor*. The equilibrium condition for the body implies $F_i = 0$, which in terms of the stress tensor can be written as

$$\sum_j \frac{\partial \sigma_{ij}}{\partial x_j} = 0. \tag{5.3}$$

In the limit of small deformations, there is a linear relation between stess and strain

$$\sigma_{ij} = \sum_{i,j,k,l} C_{ijkl}\epsilon_{kl}, \tag{5.4}$$

where C_{ijkl} is the elastic moduli tensor. The Hooke law, reported in its most general form in Eq. 5.4, usually takes a simpler form since symmetry considerations can be used to reduce the number of independent component of C_{ijkl}. In particular, for isotropic solids, we have only two independent components and the Hooke law can be written as

$$\sigma_{ij} = \frac{E}{(1+\nu)} \left(\epsilon_{ij} + \frac{\nu}{1-2\nu}\delta_{ij} \sum_{k} \epsilon_{kk} \right), \tag{5.5}$$

where E is the Young modulus and ν is the Poisson ratio. In some cases it is convenient to rewrite the strain tensor as the sum of a compressive part, involving a volume change, and a shear part, which does not. The Hooke law then becomes

$$\sigma_{ij} = K\delta_{ij} \sum_{k} \epsilon_{kk} + 2G(\epsilon_{ij} - \frac{1}{3}\delta_{ij} \sum_{k} \epsilon_{kk}), \tag{5.6}$$

where K is the bulk modulus and G is the shear modulus. These coefficients are related to the Young modulus $E \equiv \frac{9KG}{G+3K}$ and the Poisson ratio $\nu \equiv \frac{1}{2}\left(\frac{3K-2G}{3K+G}\right)$. Finally, the equation for stress equilibrium for an isotropic elastic medium can be written in terms of the displacement field $\mathbf{u}$, combining Eq. 5.5 with Eq. 5.3

$$\frac{E}{2(1+\nu)}\nabla^2\mathbf{u} + \frac{\nu E}{2(1+\nu)(1-2\nu)}\vec{\nabla}(\vec{\nabla}\mathbf{u}) = 0, \tag{5.7}$$

with appropriate boundary conditions.

5.2   Cracks in elastic media

A crack could be seen in mathematical terms as a boundary condition for Eq. 5.7, with the important complication that the precise shape of the crack is not fixed but depends on the stress field acting on it. Nevertheless, it is instructive to compute the stress field induced by a static crack. For instance, a linear crack yields a stress profile decaying from the crack tip as

$$\sigma_{ij} = \frac{K}{\sqrt{2\pi r}}f_{ij}(\theta,\phi), \tag{5.8}$$

where K is the *stress intensity factor* and f_{ij} is a universal function of the angular variables θ and ϕ. The particular values of K and f_{ij} depend on the loading conditions (i.e. the way stress is applied to the body). Typically one defines three fracture modes (see Fig. 5.1) corresponding to tensile (I), shear (II), and tearing (III) conditions, and any other load is then represented as an appropriate combination of these three modes. The presence of the square root singularity around the crack tip is an artifact of linear elasticity. Real materials cannot sustain arbitrarily large stresses and in general deformation near the crack tip is ruled by non-linear and anelastic effects. In practice it is customary to define a fracture process zone (FPZ) as the region surrounding the crack where these effects take place, while far enough from the crack one recovers linear elastic behavior.

Notice that we are interested not only in the equilibrium shape of the crack, but also in its formation and evolution. A simple argument to address this problem was proposed by Griffith in 1920 using an analogy to nucleation theory in first-order phase transitions. Considering a two-dimensional geometry, the elastic energy change associated with a circular crack of radius a is given by

$$E_{el} = -\pi\sigma^2 a^2/E, \tag{5.9}$$

where σ is the relevant component of the applied stress. Forming a crack involves the rupture of atomic bonds, a process that has an energy cost proportional to the crack surface

$$E_{surf} = 2\pi a\gamma, \tag{5.10}$$

where γ is the surface tension. Griffith noted that a crack will grow when this process leads to a decrease of the total energy $E = E_{el} + E_{surf}$. Thus the criterion for crack growth can be expressed as

$$\frac{dE}{da} = -2\pi\sigma^2 a/E + 2\pi\gamma < 0, \tag{5.11}$$

which implies that under a stress σ, cracks of size $a > a_c \equiv \gamma E/\sigma^2$ are unstable. Eq. 5.11 can also be written as $\sigma\sqrt{\pi a} > \sqrt{2E\gamma}$ and, noting that $K = \sigma\sqrt{\pi a}$ is the stress intensity factor of an infinite plate, we can reformulate the crack growth criterion in terms of a critical stress intensity factor $K_c \equiv \sqrt{2E\gamma}$, also known as the material toughness (i.e. the crack grows when $K > K_c$).

The presence of disorder has several important effects on the fracture process and in particular on the material strength. For instance, one can imagine that a random collection of microcracks of different sizes will lead to a fluctuating material toughness and result in a distribution for the failure strength. In the simple approximation of very brittle behavior, failure will be dominated by the weakest spot or the largest microcrack present in the material. Solving this *extreme value statistics* problem, one can obtain estimates of the strength distribution and quantify the size effects. As the sample gets larger it will be easier to find a weak region, leading to a smaller strength. This problem was studied in the '30s by Weibull and Gumbel and will not be discussed in more detail here. Disorder, of course, is at the root of the scaling behavior observed in AE experiments. In general most of the complexity associated with fracture can be linked to the interplay between elastic interactions and disorder, which is the focus of the remaining part of this chapter. A more extensive review of the statistical models of fracture can be found in Alava *et al.* (2006).

5.3 Acoustic emission in fracture

The characteristic crackle sound emitted by materials under stress represents a common experience for everyone. Yet this AE signal carries important information on the

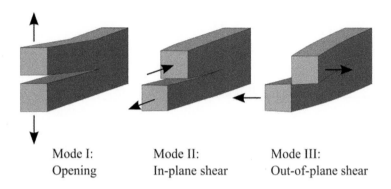

Mode I: Mode II: Mode III:
Opening   In-plane shear   Out-of-plane shear

Fig. 5.1 The three modes of fracture. Source: Image in the public domain (Wikipedia).

internal microfracturing activity taking place inside a material. Obtaining a direct image of the microfracturing process taking place inside a material is not an easy task; on the other hand, assessing the damage state of the material can be of vital importance for applications. In this respect, AE represents a simple and reliable alternative to more complicated testing methods based on thermal imaging, electromagnetic diagnostics or three-dimensional tomography to name just two. Here we will leave aside the practical aspects, although important, and focus instead on the fundamental nature of the phenomenon as a typical example of crackling noise.

In a typical AE experiment, the acoustic activity is converted into an electric signal through a piezoelectric sensor applied to the sample. In brittle materials, AE is primarily due to the release of elastic energy as a consequence of fracture, but one should bear in mind that this can also be produced by other mechanisms such as internal friction or plastic deformation, as will be discussed later on. The character of the AE signal may depend on the imposed loading conditions: One can apply a constant stress (creep test), increase the load slowly or in steps, or impose a constant deformation rate. Nevertheless, it is generally observed that the signal is composed of separated pulses with fluctuating amplitudes. As an example, we report in Fig. 5.2 two typical signals, one obtained by a slow increase of the external load in wood and the second recorded in concrete applying a few subsequent steps of stress. In the first case (Fig. 5.2a), we see that the released energy increases strongly as we approach the sample failure. The second example (Fig. 5.2b) shows instead a relaxation of the activity after the sudden increase in the stress corresponding to each step.

While the trends in the signal may vary depending on the loading conditions, one generically finds that the distribution of AE energies E follows a power-law distribution

$$P(E) \sim E^{-\tau_E}, \tag{5.12}$$

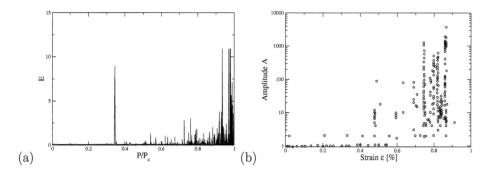

(a)   (b)

Fig. 5.2 Two examples of AE signals. (a) AE energy measured in wood by slowly increasing the load P_c, plotted as a function of P/P_c, where P_c is the failure load. (b) Aacoustic data series as a function of strain for paper samples in tensile testing. First figure: Data from Garcimartin *et al.* 1997. Second figure: Data from Salminen *et al.* 2002.

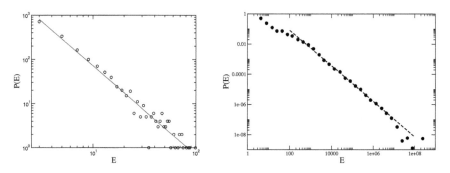

Fig. 5.3 The distribution of AE energies for wood fracture. The linear fit yields $\tau_E \simeq 2$ (left). d). The distribution of AE energies reported for paper fracture. The exponent here is $\tau_E \simeq 1.3$ (right). Source: Left image: Data from Garcimartin *et al.* (1997). Right image: Data from Salminen *et al.* (2002).

where the exponent τ_E takes different values in different experiments and materials (Petri *et al.*, 1994; Garcimartin *et al.*, 1997; Guarino *et al.*, 1998; Maes *et al.*, 1998; Guarino *et al.*, 2002; Salminen *et al.*, 2002). For instance, wood's exponent is $\tau_E = 1.51 \pm 0.05$, for fiberglass $\tau_E = 2.0 \pm 0.01$ (Garcimartin *et al.*, 1997), $\tau_E = 1.30 \pm 0.1$ for paper (Salminen *et al.*, 2002), $\tau_E = 1.5 \pm 0.1$ for cellular glass (Maes *et al.*, 1998). As an example, we report in Fig. 5.3 the AE energy distributions obtained in wood and in paper. While energy and amplitude distribution are straightforward to evaluate, in general it is difficult to determine the duration of each AE event, because of measurement system response, wave attenuation and wave dispersion, including reflections from surfaces. Thus one measures instead the time intervals t_w between successive events, scaling as

$$P(t_w) \sim t_w^{-\alpha_w}, \tag{5.13}$$

where the exponent values are typically found in the range $\alpha_w \sim 1 \ldots 1.5$.

The main problem that we have to face when trying to understand an AE signal is to reconstruct the internal process that produced it. In this respect, triangulation techniques have been devised to localize the source of each AE pulse and it thus possible, up to a certain extent, to visualize in space and time the fracturing process. A nice example of this was discussed in Guarino *et al.* (1998), where it was possible to directly observe the process of fracture localization (see Fig. 5.4). In most cases, however, we do not have such a detailed information, and require the combined analysis of the response of several transducers placed at different locations, and even in the best conditions a precise determination of the source of each event is not always possible.

In general terms, we can say that the energy content of an AE event reflects the energy released from microcrack formation and propagation. The intermittent character of the signal indicates that crack dynamics is irregular, proceeding through avalanches. In addition, the fact that the AE energy distribution is very broad, decaying as a power law, may suggest that the fracture process is related to some kind of phase transition. In order to understand to what extent this suggestion is correct, we have to look more closely at how cracks form, interact, and grow. To this end, we must first take a short detour through the theory of elasticity.

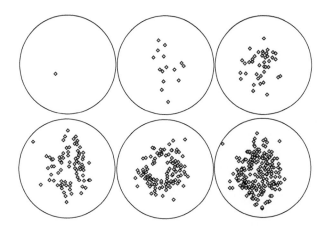

Fig. 5.4 An example of the use of AE to observe microfracture localization as the external load is increased. The final panel displays all the events from the five equidistant intervals in the previous five subplots. Source: Data from Guarino *et al.* (1998).

5.4 Dynamics of a crack line

In experiments where fracture starts from a notch, the crack front can be seen as a deformable line pushed by the external stress through a random toughness landscape, which is when the problem becomes an example of interface depinning. In these conditions, the final fracture surface is the trail left by the crack line as it advances through the medium (Bouchaud *et al.*, 1993; Daguier *et al.*, 1997). For simplicity, it is instructive to first consider the case of a planar crack front, corresponding to the experiments reported in Schmittbuhl and Maloy (1997), Delaplace *et al.* (1999), and Maloy and

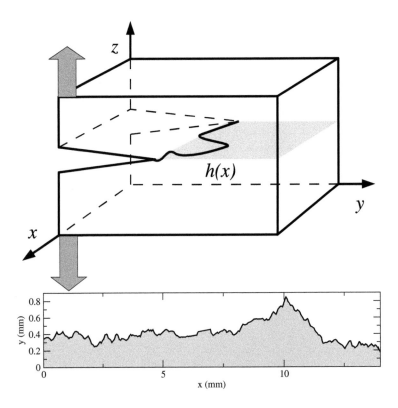

Fig. 5.5 Top: The experimental setup employed to study the dynamics of planar cracks. Bottom: The snapshots of the crack front. Source: top: Schmittbuhl and Maloy (1997, Delaplace *et al.* (1999, Maloy and Schmittbuhl (2001), bottom: Data from Santucci *et al.* (2009).

Schmittbuhl (2001). In this set of experiments, two Plexiglas plates were sandblasted and then glued together in order to create a disordered low-toughness plane where the fracture could propagate. Due to the transparency of the material it was then possible to follow the propagation of the crack as it advanced through the plane. As shown in Fig. 5.5, the crack front advances in avalanches, as would be expected from a line driven in a disordered medium. The morphology of the crack front was analyzed extensively and the roughness exponent was estimated as $\zeta = 0.63$. As for the standard interface depinning problem, one should measure the avalanche size s by analyzing the time dependence of the average crack velocity signal. Defining a suitable threshold, the avalanche size should be defined as the area under a velocity pulse. In Maloy *et al.* 2006, the authors define the size differently by looking at the map of the local waiting times. In this way, it was found that the avalanche area distribution is a power law but with an exponent $\tau_A \simeq 1.5$, independent of the crack average velocity (see Fig. 5.6 and Tallakstad *et al.* (2011)).

To understand this experiment, we can schematize the crack as a line moving on the xy plane with coordinates $(x, u(x,t))$ (Schmittbuhl *et al.* 1995; Ramanathan and

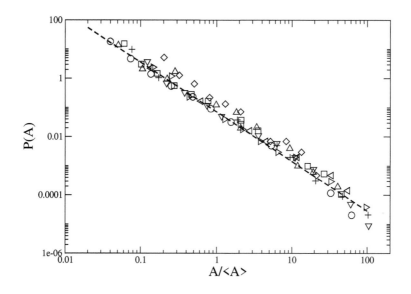

Fig. 5.6 The avalanche area distribution $P(A)$ obtained from the map of waiting times for a planar crack in the Plexiglas experiment of Maloy *et al.* (2006). The scaling exponent $\tau_A = 1.5$ is independent of the average crack velocity. Source: Data from Maloy *et al.* (2006) Fig. 3.

Fisher 1997; 1998). An equation of motion for the deformed line position is obtained by computing, from the theory of elasticity, the variations to the stress intensity factor induced by the deformation of the front. In the quasistatic scalar approximation, this is given by (Gao and Rice, 1989)

$$K(\{u(x,t)\}) = K_0 \int dx' \frac{u(x',t) - u(x,t)}{(x-x')^2}, \tag{5.14}$$

where K_0 is the stress intensity factor for a straight crack. The crack deforms because of the inhomogeneities present in the materials, which give rise to fluctuations in the local toughness $K_c(x, u(x,t))$. These ingredients can be joined together into an equation of motion of the type

$$\Gamma \frac{\partial h}{\partial t} = K_{ext} + K(\{u(x,t)\}) + K_c(x, u(x,t)), \tag{5.15}$$

where Γ is a damping term and K_{ext} is the stress intensity factor corresponding to the externally applied stress (Schmittbuhl *et al.* 1995; Ramanathan and Fisher 1997; 1998). Eq. (5.15) belongs to the general class of interface models which was discussed in Chapter 4. Due to the long-range nature of the elastic kernel, scaling as $K(q) \sim |q|$ in Fourier space, we can predict that the roughness exponent is given by $\zeta \simeq 0.35$, which is in disagreement with early experiments (Schmittbuhl and Maloy, 1997; Delaplace *et al.*, 1999; Maloy and Schmittbuhl, 2001), but in agreement with more recent experiments, suggesting that the high value of the roughness exponent (i.e. $\zeta = 0.6$) is due to small microcrack coalescence (Santucci *et al.*, 2009) and crosses over to $\zeta = 0.35$ at larger length scales (see Fig. 5.7). The validity of Eq. 5.15 to describe planar crack front propagation is also confirmed by numerical simulations yielding $\tau = 1.25$ (Bonamy *et al.*, 2008), as expected from the depinning transition (see Table 4.3).

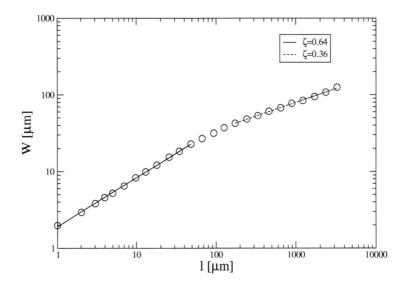

Fig. 5.7 The width W of the planar crack as a function of the scale l. The data show a crossover scaling from $\zeta = 0.6$, consistent with previous experiments, to $\zeta = 0.35$ as expected from the theory. The crossover length at $l_c = 100\mu$m corresponds to the scale of the microstructure. Source: Data from Santucci *et al.* 2009.

The exponent τ_A is different from τ because it relates to the distribution of *connected* clusters. For short-range interfaces, avalanches are connected, but this is not true when long-range interactions are present (Laurson *et al.*, 2010). In this case, an avalanche is composed of a set of disconnected clusters which are distributed as a power law with exponent τ_A. It is possible to derive a scaling relation between τ and τ_A (Le Priol *et al.*, 2021):

$$\tau_A = 2\tau - 1. \tag{5.16}$$

Notice that the same scaling relation was derived earlier (Laurson *et al.*, 2010) but some assumptions made in the derivation were later proven not to be correct in general (Le Priol *et al.*, 2021). Inserting $\tau = 1.25$ in Eq. 5.16, we obtain $\tau_A = 1.5$ which is in agreement with numerical simulations (Le Priol *et al.*, 2021; Laurson *et al.*, 2010) and experiments (Laurson *et al.*, 2010).

The more general problem of three-dimensional crack propagation has also been formulated as a depinning problem, but similar difficulties arise (Ramanathan *et al.*, 1997). For this problem, a large body of experimental literature points toward a universal roughness exponent $\zeta \simeq 0.78$, a value measured at different scales and in widely different materials. If the crack can propagate out of plane, we have to introduce an additional component $f(x, y)$ to describe these displacements. The in-plane component $h(x)$ still follows Eq. (4.7), but is now coupled to the out-of-plane component

$f(x, y)$ (Ramanathan *et al.*, 1997). The resulting roughness exponent is predicted to depend on the fracture mode: In mode I the out-of-plane roughness is only logarithmic, while an exponent $\zeta = 1/2$ is found in mode III. Both results are quite far from the experimental results $\zeta \simeq 0.78$. Experimental evidence shows, however, that the picture is more complicated: The above scaling is valid at small scales, while at large scales one observes a new regime with $\zeta \simeq 0.4$ attributed to crack line depinning (Ponson *et al.*, 2006; Bonamy *et al.*, 2006). In this respect, experiments on amorphous silica have shown that for three-dimensional cracks the roughness correlation length ξ is of order of the size of the fracture process zone (FPZ) ξ_{FPZ} (Célarié *et al.*, 2003; Marliére *et al.*, 2003; Prades *et al.*, 2004). In this region around the crack tip, linear elasticity does not apply and a well-defined elastic crack front is not present.

Bonamy *et al.* (2006) propose a modification of the coupled equation for the crack front line that ultimately gives rise to an equation in the same universality class as Eq. 5.15. Hence, one would expect a roughness exponent of $\zeta \simeq 0.35$, which is in good agreement with the experiments. The small scale regime described by $\zeta = 0.78$, observed in different materials and measured over several decades of scaling, would then also be due to microcrack coalescence inside the FPZ, but a quantitative explanation is lacking. To solve this issue, a possible theoretical path sees instead crack dynamics as the coalescence of local microcracks, abandoning the concept of a well-defined crack line. A similar mechanism has also been posed for planar crack propagation (Zapperi *et al.*, 2000; Aström *et al.*, 2000; Schmittbuhl *et al.*, 2003; Bouchbinder *et al.*, 2004). In the following, we discuss models of interacting microcracks in disordered media, starting from the simplest, the fiber bundle model , for which an analytic treatment is possible. Clearly these models are more appropriate to describing materials with strong disorder than the case of a single crack starting from a notch. Nevertheless, one can still impose a geometry with a pre-existing crack in order to validate line models.

5.5 Fiber bundle models

Fiber bundle models (FBM), originally introduced to study the failure of fibrous systems, provide a simplified picture of the fracture of a disordered material (Pierce, 1926; Daniels, 1945). The model consists of a set of brittle fibers loaded in parallel with random failure thresholds. When the load on a fiber overcomes its threshold, the fiber fails and the load is redistributed to the other fibers according to a prescribed rule. The simplest possibility is the case of an equal load sharing (ELS), in which each intact fiber carries the same fraction of the load. This case represents a sort of mean-field approximation and allows for a complete analytical treatment (Daniels, 1945; Hemmer and Hansen, 1992; Hansen and Hemmer, 1994; Sornette, 1992*b*; Kloster *et al.*, 1997). At the other extreme lies the local load sharing (LLS) model where the load of a failed fiber is redistributed to the intact neighboring fibers (Harlow and Phoenix, 1978; Smith, 1980; Smith and Phoenix, 1981; Phoenix and Ra, 1992; Curtin, 1991; Curtin, 1993; Zhou and Curtin, 1995; Leath and Duxbury, 1994; Zhang and Ding, 1996; Beyerlein and Phoenix, 1996; Phoenix *et al.*, 1997).

These simplified models serve as a basis for more realistic damage models such as the micromechanical models of fiber-reinforced composites which take into account stress localization (Harlow and Phoenix, 1978; Smith and Phoenix, 1981), the effect of

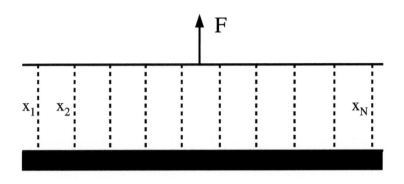

Fig. 5.8 In the fiber bundle model , N fibers in parallel are subjected to a load F and have random failure thresholds x_i. When a fiber fails, its load is transferred to the others according to a given rule.

matrix material between fibers (Harlow and Phoenix, 1978; Smith and Phoenix, 1981; Phoenix and Ra, 1992; Curtin, 1991; Curtin, 1993; Zhou and Curtin, 1995; Phoenix *et al.*, 1997), and possible non-linear behavior of fibers (Krajcinovic and Silva, 1982). Other generalizations of the fiber bundle model include viscoelastic couplings (Fabeny and Curtin, 1996; Hidalgo *et al.*, 2002; Kun *et al.*, 2003), continuous damage (Kun *et al.*, 2000), and thermally activated fracture (Roux, 2000; Scorretti *et al.*, 2001; Politi *et al.*, 2002).

While FBM has been used to model several aspects of the failure of fibrous materials, we concentrate here on the avalanche behavior: The failure of a fiber in general will induce subsequent failures as a consequence of the load transfer. We consider first the case of ELS fiber bundles, in which N fibers of unitary Young modulus $E = 1$ are subject to a uniaxial load F. Each fiber i obeys the linear elastic equation up to a critical load x_i, which is randomly distributed according to a distribution $p(x)$ (see Fig. 5.8). When the load on a fiber exceeds x_i, the fiber is removed. Due to the ELS rule, when n fibers are present each of them carries a load $F_i = F/n$ and consequently a strain $\epsilon = F/n$.

The average response for ELS fiber bundles can easily be obtained from a self-consistent argument. At a given load F, the number of intact fibers is given by

$$n = N \left(1 - \int_0^{F/n} p(x)dx \right) = N\epsilon(1 - P(\epsilon)), \qquad (5.17)$$

where $P(x)$ is the cumulative distribution obtained from $p(x)$ and $\epsilon = F/n$ is the strain. As a simple illustration, we consider a uniform distribution in $[0,1]$: In this case the fraction of intact fibers $\rho \equiv n/N$ is given by

$$\rho = (1 + \sqrt{1 - 4f})/2. \qquad (5.18)$$

This equation shows that as the load is increased, ρ decreases up to $f_c = 1/4$ at which $\rho = \rho_c = 1/2$. For larger loads Eq. 5.18 displays no real solution, indicating the onset of catastrophic failure. It is interesting to rewrite Eq. 5.18 as

$$\rho = \rho_c + A(f_c - f)^{1/2}, \tag{5.19}$$

with $\rho_c = 1/2$, $f_c = 1/4$, and $A = 1$. In fact, this form is generically valid for most distributions of $p(x)$. To see this, we rewrite Eq. (5.17) as $f = x(1 - P(x))$ where $x \equiv f/\rho$ is the load per fiber. Failure corresponds to the maximum x_c of the left-hand side, after that there is no solution for $x(f)$. Expanding close to the maximum we obtain $f \simeq f_c + B(x - x_c)^2$, which then leads to Eq. 5.19 for a wide class of distributions (da Silveira, 1998).

Eq. (5.19) also implies that the average rate of bond failures, or in other words the *average avalanche size*, diverges as f_c is approached:

$$\frac{d\rho}{df} \sim (f_c - f)^{-1/2}. \tag{5.20}$$

The avalanche distribution can be computed analytically considering the sequence of external loads $\{F_k\}$ corresponding to failure events. If we order the thresholds $\{x_k\}$, the kth failure event will occur at a load $F_k = (N - k + 1)x_k$, and on average we have

$$\langle F_{k+1} - F_k \rangle = (1 - P(x_k))p(x_k). \tag{5.21}$$

Close to the global failure point x_c, we expect that $\langle F_{k+1} - F_k \rangle \sim x_c - x_k$ and that $\langle (F_{k+1} - F_k)^2 \rangle$ goes to a constant as $x \to x_c$. Hence, the external load $\{F_k\}$ performs a biased random walk, with a bias vanishing as $x \to x_c$. For a given value of the load, an avalanche is defined by the return of the variable F_k to its initial value. Thus the avalanche size distribution is just the first return time distribution of a biased random walk:

$$D(s, f) \sim s^{-3/2} \exp(-s/s_0), \tag{5.22}$$

where the cutoff $s_0 \sim (x_c - x)^2 \sim (f_c - f)^{-1}$ diverges as $f \to f_c$ is the limit $N \to \infty$. Eq. (5.22) refers to the avalanches at a fixed value of the load f. If we instead consider the integrated avalanche distribution obtained when the load is increased from zero to complete failure, we obtain

$$D_{int}(s) \sim s^{-5/2}. \tag{5.23}$$

While the ELS rule treats the elastic interaction at the mean-field level without any stress enhancement, in LLS the load is transferred only around the crack. In the simplest one-dimensional version of the model, when a fiber fails, its load is redistributed to the neighboring intact fibers. Thus the load on a fiber is given by $f_i = f(1 + k/2)$, where k is the number of failed fibers that are nearest neighbors of the fiber i and $f = F/N$ is the external load (Harlow and Phoenix, 1978). Even for this apparently simple one-dimensional model a closed form solution is not available, but several results are known from numerical simulations, exact enumeration methods, or approximate analytic calculations. In particular, the average bundle strength decreases as the bundle size grows as

$$f_c \sim 1/\log(N), \tag{5.24}$$

so that an infinitely large bundle has zero strength. Nevertheless, one can study the avalanche distribution for bundles of finite size. Numerical simulations show that the integrated avalanche distribution is well approximated by

$$D_{int}(s) \sim s^{-4.5} \exp(-s/s_0), \tag{5.25}$$

where s_0 is independent of N. We may conclude that no true scaling is observed when interactions are short-ranged.

5.6   Lattice models for fracture

5.6.1   The random fuse model

The simplest way to study the effect of elastic interactions in the fracture of a disordered medium is provided by the random fuse model (RFM) (de Arcangelis *et al.*, 1985; Herrmann and Roux, 1990), where a set of electrical elements (fuses) are placed into a lattice (see Fig. 5.9). A voltage drop is imposed between two sides of the lattice and the current in each bond is obtained solving the Kirchhoff equations, which depend on the local conductivities σ_{ij}. When the local current overcomes a threshold current i_c, the fuse fails irreversibly and its conductivity is set to zero. When this happens one recomputes the currents in the lattice and checks whether some other fuses should fail. When all fuses are below threshold, the external voltage is raised and the process is repeated.

Disorder is introduced in the RFM by extracting the thresholds $i_{c,ij}$ of each fuse from a probability distribution $p(x)$, or by removing a fraction p of the links at the beginning of the simulation. One can then tune the amount of disorder present in the system by changing the distribution or the dilution ratio. The crossover from weak to strong disorder is mostly studied in the framework of the first case using a uniform distribution in $[1 - R, 1 + R]$ (Kahng *et al.*, 1988) or a power law distribution $p(x) \sim x^{-1+\beta}$ in $[0, 1]$ (Hansen *et al.*, 1991) and cumulative distribution $F(x) = x^\beta$. The first case ranges from no disorder ($R = 0$) to strong disorder ($R = 1$), while in the second case one can study extremely strong disorder, since for $\beta \to 0$ the distribution becomes non-normalizable. The case $R = 1$ is probably the most studied in the literature. Alternatively it is also possible in principle to connect the disorder in the model to a realistic material microstructure by prescribing the fuse properties based on the morphology of the material microstructure.

It is important to remark that the Kirchhoff equations can be considered as a discretization of the continuum Laplace equation. While this represents a strong simplification of the complete elastic problem, it formally corresponds to Eq. 5.7 in the limit of an antiplanar shear deformation. Nevertheless, the interest of the model does not reside in its realism but mostly in its simplicity. More realistic models can be devised, as we will discuss in what follows, but the main features of the avalanche distributions appear to be independent of this.

5.6.2   Tensorial models

Following the general strategy of the RFM, several models with more realistic elastic interactions (see section 5.1) have been proposed in the past. The simplest possibility is provided by central force systems, where nodes are connected by elastic springs (Hansen *et al.*, 1989; Arbabi and Sahimi, 1993; Sahimi and Arbabi, 1993; Nukala *et al.*, 2005). The random spring model (RSM) is defined by the Hamiltonian

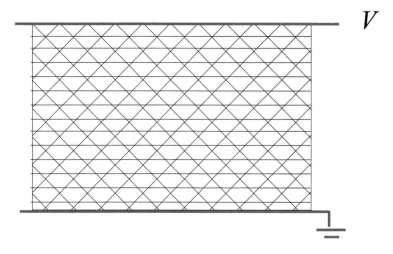

Fig. 5.9 Random-threshold fuse network with triangular lattice topology. Each of the bonds in the network is a fuse with unit electrical conductance and a breaking threshold randomly assigned based on a probability distribution (uniform). The behavior of the fuse is linear up to the breaking threshold. Periodic boundary conditions are applied in the horizontal direction and a unit voltage difference, $V = 1$, applied between the top and bottom of lattice system bus bars. As the current I flowing through the lattice network is increased, the fuses will burn out one by one.

$$H = \sum_{ij} \frac{K}{2} (\mathbf{u}_i - \mathbf{u}_j)^2, \qquad (5.26)$$

where $\mathbf{u}_i$ is the displacement of node i and K the spring constant. The elastic equilibrium is obtained by minimizing Eq. (5.26) and disorder is introduced in the standard way by dilution, random threshold, or random elasticity. Despite the fact that the RSM suffers from the caveats associated with the presence of soft modes, it has still been considered in the context of fracture. In terms of continuum elasticity, the discrete triangular central-force lattice model represents an isotropic elastic medium with a fixed Poisson's ratio of $1/3$ in two dimensions, and $1/4$ in three dimensions.

A similar, slightly more complicated model is the Born model, defined by the Hamiltonian

$$H = \sum_i \sum_j \alpha((\mathbf{u_i} - \mathbf{u_j}) \cdot \mathbf{x}_{ij,\|})^2 + \beta((\mathbf{u_i} - \mathbf{u_j}) \cdot \mathbf{x}_{ij,\perp})^2. \qquad (5.27)$$

In this case, there is a primitive competition between stretching and bending the "bond" between two nodes on a lattice and the components parallel and perpendicular to it ($\mathbf{x}_{ij} = \mathbf{x}_j - \mathbf{x}_i$). Note that in the limit $\alpha = \beta$, the Born model reduces to the RFM, but does not consider the energy cost of local rigid rotations (see Feng and Sen 1984; Hassold and Srolovitz 1989).

Finally, a more realistic treatment of elastic interactions is found in beam and bond-bending models. The first model can be seen as a lattice of massless particles connected by a beam to their nearest neighbors (Roux and Guyon, 1985). Each beam is capable of sustaining longitudinal (F) and shear forces (S), and a bending moment (M). The bond-bending model is instead a generalization of the central force model where one includes the energy costs of local rotations of the bonds. The inclusion of bond-bending terms into a triangular lattice system results in a conventional two-dimensional isotropic elastic medium with varying Poisson's ratio values (Monette and Anderson, 1994). On the other hand, the triangular beam model lattice system represents a Cosserat continuum.

In general, tensorial fracture models require the choice of a fracture criterion to be used, which in principle should depend on the combination of local torque, and longitudinal and shear forces in beam models, or on central and angular forces in the bond-bending model.

5.6.3 Avalanche distributions

As for the fiber bundle model s discussed in section 5.5, the fracture of lattice models proceeds by avalanches whose sizes grow as the external load is increased. The main problem is to relate the distributions obtained from simulations to the AE statistics recorded in experiments, since in the models the avalanche size s is defined by the number of broken bonds that fail without any further increase in the applied load. This avalanche size definition is convenient from the numerical viewpoint, but it is not equivalent to the acoustic emission energy, so some additional assumption is needed to relate the two distributions, as we will discuss in the following.

The most extensive avalanche statistics for lattice models have been obtained for the RFM in two (Hansen and Hemmer, 1994; Zapperi *et al.*, 1997*a*; Zapperi *et al.*, 1999; Zapperi *et al.*, 2005*c*) and three (Räisänen *et al.*, 1998; Zapperi *et al.*, 2005*b*) dimensions, but results have been reported for tensorial models as well. The first observation is that if we simply record the distribution considering all the avalanches, we would observe a power law decay culminating with a peak at large avalanche sizes. This peak is due to the last catastrophic event which can thus be considered as an outlier and analyzed separately, avoiding possible bias in the exponent estimate. When the last avalanche is removed from the distribution the peak disappears and one is left with a power law (see Zapperi *et al.* 2005*c*). On the other hand, the distribution of the last avalanche is Gaussian (Caldarelli *et al.*, 1999; Zapperi *et al.*, 2005*c*; Nukala *et al.*, 2005).

Next, as discussed in section 5.5, one should distinguish between distributions sampled over the entire load history, and those recorded at a given load. In the first case, the integrated distribution has been described by a scaling form (Zapperi *et al.*, 2005*c*)

$$P(s, L) = s^{-\tau_{int}} g(s/L^D), \tag{5.28}$$

where D represents the fractal dimension of the avalanches. To take into account the different lattice geometries, it is convenient to express scaling plots in terms of N_{el} rather than L:

$$P(s, N_{el}) = s^{-\tau_{int}} g(s/N_{el}^{D/2}). \tag{5.29}$$

A summary of the results for two-dimensional lattices is reported in Fig. 5.10, showing that in the RFM $\tau_{int} > 5/2$, with variations depending on the lattice topology, while the result reported for the RSM is instead very close to $\tau_{int} = 5/2$. As for the avalanche fractal dimension, D varies between 1.10 and 1.18, indicating almost linear crack (which would correspond to $D = 1$). Three-dimensional results are plotted in Fig. 5.11 for the RFM (Zapperi *et al.*, 2005*b*). Again, the agreement with FBM results is quite remarkable.

The distribution of avalanche sizes sampled at different values of the current I is obtained by normalizing the current by its peak value I_c and dividing the $I^* = I/I_c$ axis into several bins. The distribution follows a law of the type

$$p(s, I^*) = s^{-\tau} \exp(-s/s_0), \tag{5.30}$$

where the cutoff s_0 depends on I^* and L, and can be conjectured to scale as

$$s_0 \sim \frac{L^D}{(1 - I^*)^{1/\sigma} L^D + C}, \tag{5.31}$$

where C is a constant. Integrating Eq. (5.30) we obtain

$$P(s, L) \sim s^{-(\tau+\sigma)} \exp[-sC/L^D], \tag{5.32}$$

which implies $\tau_{int} = \tau + \sigma$. We recall here that in the FBM $\tau = 3/2$ and $\sigma = 1$ (see section 5.5). The best fit for the scaling of the distribution moments as a function of I and L (Eq. 5.31), and a direct fit on the curves, suggest $\tau \approx 1.9 - 2$, $\sigma \approx 1.3 - 1.4$ in $d = 2$ and $\tau \approx 1.5$, $\sigma \approx 1$ in $d = 3$.

In summary, the avalanche distribution for RFM is qualitatively close to the prediction of the FBM (i.e. mean-field theory). Apparently non-universal (lattice type dependent) corrections are present in $d = 2$, but the problem disappears in $d = 3$. This might be expected, building on the analogy with critical phenomena, where exponents are expected to approach the mean-field values as the dimension is increased. Simulations of other models have been reported in the literature: The distribution obtained in Nukala *et al.* (2005) for the RSM is very close to the FBM result (see Fig. 5.10). The avalanche distribution has also been reported for the Born model ($\tau_{int} \approx 2$ (Caldarelli *et al.*, 1996)), the bond-bending model ($\tau_{int} \approx 2.5$ (Zapperi *et al.*, 1999)), and the continuous-damage model ($\tau \approx 1.3$ (Zapperi *et al.*, 1997*b*)).

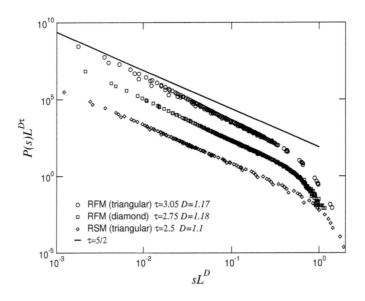

Fig. 5.10 A summary of two-dimensional results for avalanche distributions integrated along the entire loading curve. The results are for the RFM on triangular and diamond lattices (Zapperi *et al.*, 2005*c*) and for the RSM (Nukala *et al.*, 2005). Data for different lattice sizes have been collapsed according to Eq. 5.29. A line with exponent $\tau = 5/2$ is reported for reference.

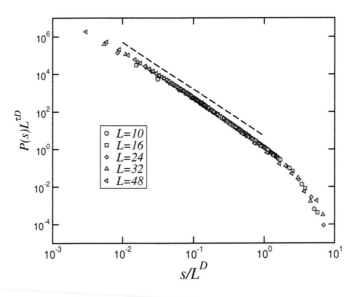

Fig. 5.11 Data collapse of the integrated avalanche size distributions for the three-dimensional RFM. The exponents used for the collapse are $\tau = 2.5$ and $D = 1.5$. A line with a slope $\tau = 5/2$ is reported for reference. Source: Data from Zapperi *et al.* (2005*c*).

To conclude, we discuss the issue of the relation between the AE distribution and and the avalanche size distribution. These two are, contrary to usual indications in the literature, not the same. The same holds even more clearly for the acoustic amplitude distribution. Acoustic waves were incorporated in a scalar model in Minozzi *et al.* (2003), and the AE energy distribution was measured, with an exponent $\tau_E \simeq 1.7$. This exponent can be related to τ_{int} if we assume that the typical released AE energy of an avalanche of size s scales as $E \sim s^\delta$, with $\delta = 2$ (i.e. the scaling relation is $\tau_{int} = 1 + \delta(\tau_E - 1)$). This is verified in the model of Minozzi *et al.* (2003), but also in the RFM defining E as the dissipated electric energy during one event, which seems to agree (Salminen *et al.*, 2002).

5.7 Fracture as a phase transition

The interpretation of fracture as a phase transition or as a critical phenomenon has been discussed many times in the literature but the issue still remains controversial. There is a clear analogy between the classical theory of nucleation and the Griffith theory discussed in section 5.2. Several authors have pointed out that an homogeneous solid under stress is in a metastable state, which becomes unstable when the external stress reaches a spinodal-like point (Rundle and Klein, 1989; Selinger *et al.*, 1991*a*; Selinger *et al.*, 1991*b*; Wang *et al.*, 1991; Golubovic and Feng, 1991; Golubovic and Pedrera, 1995; Buchel and Sethna, 1997), but this reasoning mainly applied to thermally activated fracture in homogeneous media. The presence of disorder has a strong effect on the nucleation process (Selinger *et al.*, 1991*a*; Golubovic and Feng, 1991; Golubovic and Pedrera, 1995), providing in some cases a disorder-induced effective temperature (Arndt and Nattermann, 2001). On the other hand, in several instances fracture is not thermally activated but stress driven, making the nucleation picture inapproprate. A different path of investigation stems from the scaling laws observed in fracture morphologies and AE distributions, suggestive of the presence of an underlying critical point (Hansen and Hemmer, 1994; Zapperi *et al.*, 1997*a*; Zapperi *et al.*, 1999; Andersen *et al.*, 1997; Barthelemy *et al.*, 2002; Toussaint and Pride, 2005).

The lattice models for fracture discussed in section 5.6 provide an instructive playground to elaborate on these analogies, since it is sometimes possible to map them to spin or Ising type models where phase-transitions are well understood (Zapperi *et al.*, 1997*a*; Zapperi *et al.*, 1999; Barthelemy *et al.*, 2002; Toussaint and Pride, 2005). We consider first the fiber bundle model , since the nature of a possible phase transition can be understood owing to the exact solution of the model (Zapperi *et al.* 1997*a*, (1999)). We can transform the FBM into an Ising model by assigning to each fiber a spin variable s_i, whose value depends on whether the fiber is intact ($s = -1$) or broken ($s = 1$). Using these variables the failure process can be mapped to a zero temperature spin flip dynamics (see Eq. 3.7 in section 3.3), with an effective field f_i given by

$$f_i = F_i(\{s_j\}, F) - x_i. \tag{5.33}$$

Here F_i is the load on the fiber i, which depends on the external load F and on the state of the other fibers through the load transfer rule of the model. As in the RFIM, each "spin" follows the sign of the local field and the fiber breaks when $F_i > x_i$.

In the case of ELS fiber bundles, the load on each fiber is given by $F_i = 2F/(N - \sum_j s_j)$ while for LLS F_i depends on the state of the neighboring fibers: If a fiber is close to a crack of length k, then $F_i = F(1 + k/2)$. The crucial difference between the FBM and the RFIM is that the load transferred to a fiber due to the failure of another fiber is increasing with the external load F, while the effective field of a spin always increases by the same amount $2J$ independently on the applied field. To be more precise, after breaking a fiber, the effective field of the others increases by a quantity $\Delta F = F/(n(n-1))$ for ELS and $\Delta F = kF/2$ for LLS. As we approach global failure, the load transferred becomes very large, because ΔF simply diverges as $n \to 0$ for ELS or increases as the cluster size k for LLS. Thus interactions always prevail against disorder and, even for very strong (but finite) disorder, it is not possible to fracture the bundle by a series of small avalanches as in the RFIM: Eventually the load transferred is so large that a catastrophic avalanche is nucleated. In this sense the transition is always of first-order type. The scaling observed in the ELS models is the one associated with a spinodal instability, similar to what is found in the low-disorder phase of the RFIM in mean-field theory. When interactions are local, such as for the LLS model, the spinodal line cannot be reached and the transition is even sharper, without large precursors.

While a phase-transition interpretation of fracture in fiber bundle model s poses relatively few problems, although in several instances the spinodal scaling was mistakenly considered as a signature of a second-order transition, in general the situation is more complicated. Attempts were made in the past to map fracture models into spin models, taking into account the long-range nature of the elastic interactions. In practice, the local stress in Eq. 5.33 is expressed in terms of a suitable lattice Green function G_{ij}, expressing the load transferred to j after a bond j has been broken. Consider for instance the random-fuse model: When only a few bonds are broken, G_{ij} is well approximated by a dipolar field, obtained from the solution of the electric problem in the undamaged lattice. As we approach the failure point, however, the local current cannot be expressed as a superposition of homogeneous Green functions and finding the correct G_{ij} amounts to solving the full problem (i.e. the Kirchhoff equations in the damaged lattice). In other words, we can rigorously map fracture models into a dipolar spin model, but only in the dilute limit which is not relevant for fracture. Extrapolating the results of spin models outside their regime of validity leads to inconsistent results, because the stress amplification at the crack tips is not properly captured by the dilute Green function (Barthelemy *et al.*, 2002; Toussaint and Pride, 2005).

A comprehensive picture of criticality in fracture can be obtained considering the fuse model and interpolating between strong and weak disorder. If we model disorder with a cumulative threshold distribution $F(x) = x^\beta$ with $x \in [0, 1]$, in the limit $\beta \to 0$, fuses break one by one and the fracture process is equivalent to percolation (Roux *et al.*, 1988) while in the limit $\beta \to \infty$ disorder is infinitesimal and fracture occurs as a nucleation process. At intermediate disorder, unlike typical first-order transitions, crack nucleation is preceded by avalanches with power-law distributions and mean-field exponents (Zapperi *et al.*, 1999). It can be argued that stress concentration at the crack tip is a relevant perturbation to the percolation critical point in the large disorder limit ($\beta > 0$) (Shekhawat *et al.*, 2013). Hence, percolation-like behavior is only a finite-size effect that at larger sizes crosses over to the mean-field avalanche behavior and then eventually to crack nucleation (Shekhawat *et al.*, 2013). This results in a *finite size criticality* in which the critical behavior is only apparent at finite sizes is illustrated in the phase diagram in Fig. 5.12.

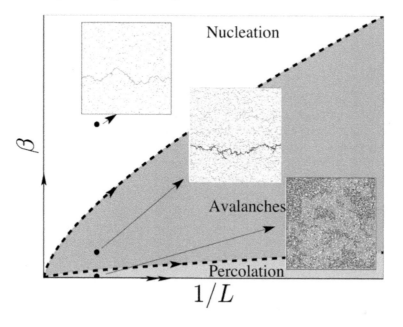

Fig. 5.12 Phase diagram for brittle fracture in disordered media according to Shekhawat *et al.* (2013). Disorder decreases along the β axis. Nucleation governs the behavior for small disorder or long length scales. percolation is characteristic of the large disorder regime, while the crossover region exhibits interesting critical behavior in the form of scale-free distributions of avalanche sizes. The topology of the fractured samples evolves from percolation-like damage for large disorder to well-defined sharp cracks in the nucleated regime. The phase boundaries are quantitatively somewhat arbitrary, and are set at the value of the scaling variable $\beta L^{1/\nu_f}$ at which the second moment of the avalanche size becomes half of its peak value (for the avalanche phase); the boundary of the percolation phase is found in a similar way. Source: Adapted from Shekhawat *et al.* (2013).

6

Plasticity

Contrary to fracture, plastic deformation is usually associatd with smooth flow and one would not expect to observe any avalanche behavior. Nevertheless, we have known about plastic instabilities with strong and widely fluctuating deformation jumps for a long time. These are typically attributed to the interplay between dislocation and diffusing solute atoms. Experiments have shown that in plastically deformed single-crystals, acoustic emission displays power-law amplitude distributions. In the presence of solute, dislocations display a depinning transition in response to external stress. The remarkable nature of the mutual dislocation interactions is also responsible for a similar jamming transition even without intrinsic pinning. In this chapter, we discuss avalanche in crystal plasticity, dislocation pinning and jamming, and finally review the properties of plastic deformation in amorphous materials and glasses.

6.1 Continuum plasticity

Plasticity can be defined as an irreversible deformation of a solid body that persists after the sample has been unloaded. As for the case of fracture, a solid undergoes plastic deformation when the stress σ is larger than a given threshold, known as the yield stress. Solids that prior to fracture accumulate some amount of plastic deformation are defined as ductile, as it is the case for most metals. Notice that in Chapter 5 we mostly considered the case of brittle materials, where fracture occurs when the solid is still elastic, or quasi-brittle ones in which a considerable amount of damage is accumulated before final rupture. Plastic deformation is a quite different phenomenon since it is not associated with any form of microcracking but rather with relative displacements of adjacent lattice planes. These rearrangements should preserve the crystal lattice structure and this in turn has an important effect on the structure of the plastic strain tensor as we will detail here.

In general, plastic deformation can be described as an additional component to the strain tensor (see Eq. 5.1) that can be decomposed as

$$\epsilon_{ij} = \epsilon_{ij}^{el} + \epsilon_{ij}^{p},\tag{6.1}$$

where ϵ_{ij}^{el} is the usual elastic strain tensor, related to the stress tensor by the Hooke law Eq. 5.4, and $\epsilon_{ij}p$ is the accumulated plastic strain tensor. As mentioned above, plastic deformation must occur by the shear of adjacent lattice planes by a lattice vector b_i, the Burgers vector, contained in the plane. The slip directions are then defined by the Burgers vector b_i and the vector normal to the plane n_i where deformation is occurring, which usually correspond with the most close-packed crystallographic planes. As a consequence of this observation the plastic strain tensor can be decomposed into the sum of the scalar shear strains $\gamma^{(\alpha)}$ on the different slip systems α

$$\epsilon_{ij}^p = \sum_\alpha \gamma^{(\alpha)} M_{ij}^{(\alpha)}, \tag{6.2}$$

where the projection tensor can be written explicitly as $M_{ij} = (n_i b_j + n_j b_i)/2b$. The inner product of the stress tensor with the same projection tensor yields the resolved shear stress $\sigma^{(\alpha)} = M_{ij}^{(\alpha)} \sigma_{ij}$ acting on the slip system. While this decomposition is convenient in general, here for simplicity we will confine most of our discussion to the case of a single slip system. The great advantage will be that we deal only with purely scalar quantities.

The plastic response of a solid is often characterized by a constitutive law, relating the plastic strain to the stress. The simplest example is probably the case of *ideal plasticity*: The solid responds elastically up to the yield stress σ_y and after that deformation proceeds indefinitely without the need to increase the stress further. When the stress is decreased below σ_y the accumulated plastic shear stress remains stored in the system as the solid unloads elastically (see Fig. 6.1). A generalization of this constitutive law involves work hardening: After the yield stress plastic deformation requires a linearly increasing stress to occur. The slope θ of the resulting plastic stress-strain curve is known as the *hardening coefficient* and should not be confused with an elastic modulus since it describes an irreversible deformation. Again, on unloading the accumulated plastic strain is stored into the system. More complicated plastic constitutive laws are encountered and used to describe the behaviors of materials. In general, they can be written in the form of a *flow stress* σ_f, the stress required to deform the sample for a given plastic strain γ and/or strain rate $\dot\gamma$:

$$\sigma_f = \sigma_y + f(\gamma, \dot\gamma) \quad \text{for} \ |\sigma| > \sigma_y, \tag{6.3}$$

where σ_y is the value for which plasticity starts and f is a constitutive function. For instance, f is zero for ideal plasticity, it is a linear function of γ in the case of strain hardening, and for viscoplasticity it is a function of $\dot\gamma$. Such a compact description is very general and appealing, but it represents only an idealization for real materials. In fact, creep deformation occurs even at low stress and it can sometimes be difficult to define unambiguously the value of the yield stress. We encounter a similar problem when we try to define a depinning threshold in the presence of noise or thermal activation. Thus for the purpose of the present discussion we will assume that yield and flow stress have well-defined unambiguous values.

To fix the ideas, it is instructive to consider a viscoplastic constitutive relation

$$\dot\gamma = C[\sigma_e - \sigma_f(\gamma)] \quad \text{for} \ \sigma_e > \sigma_y, \tag{6.4}$$

where C is a constant and the flow stress σ_f is a function of γ only. Stable plastic flow occurs for $C > 0$ and $\theta \equiv \frac{d\sigma_f}{d\gamma} > 0$. In some experimental conditions, however, these inequalities do not hold and one talks about strain softening ($\theta < 0$), or strain-rate softening ($C < 0$). In both cases, we expect plastic instabilities and the stress–strain curve will display macroscopic jumps, analogous to the stick–slip motion of a frictional

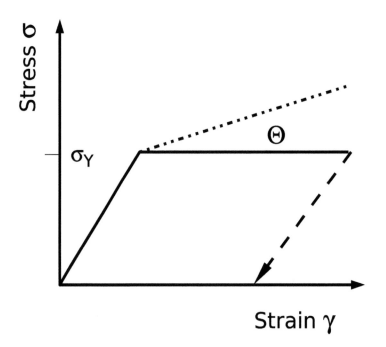

Fig. 6.1 Plastic constitutive laws. In ideal plasticity (solid line) the stress increases elastically until the yield stress and then remains constant. Upon unloading the stress follows the dashed line. When work hardening is present, the flat profile grows as in the dotted line, defining the coefficient θ. Notice that this is an irreversible deformation not to be confused with an elastic deformation.

body. This effect is known by the names of Portevin and Le Chatelier (Zaiser and Hähner, 1997; Ananthakrishna *et al.*, 1999; Ananthakrishna, 2007).

6.2 Dislocations

plastic deformation is always a localized process, since it occurs along some particular lattice planes. In fact the plastic strain is carried by quanta, the dislocations (Hirth and Lothe, 1992). These are topological defects of the lattice corresponding to the boundary of a slipped area and are characterized by their Burgers vector **b**, equivalent to a topological charge. In the simplest case the slipped area is a semiplane and the dislocation is a straight line, but other possibilities occur: The line can be deformed or the dislocation can form a closed loops. For topological reasons the dislocation can not end inside the lattice so it must either form a loop or intersect the sample boundaries.

A dislocation line is completely defined by the line tangent vector **t** and the Burgers vector **b**. When the two vectors are parallel to each other, we have a *screw* dislocation and when they are perpendicular, an *edge* one (see Fig. 6.2). In general, however, a dislocation will have a mixed character since while **b** is a constant all along the dislocation line, **t** changes if the line is not straight. In the presence of an elastic stress σ_{ij}, dislocations experience a force per unit length which is given by the Peach–Koheler

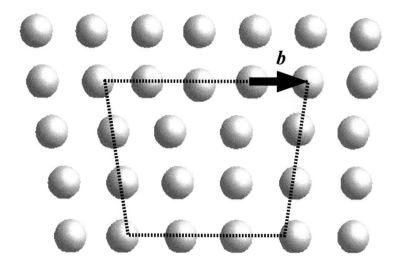

Fig. 6.2 An edge dislocation in the square lattice. A circuit around the dislocation following the lattice directions does not close, defining the Burgers vector **b**. In a cubic lattice the dislocation would extend as a line going out of the plane.

expression

$$F_i^{(PK)} = \sum_{jkl} \sigma_{ij}\epsilon_{jkl}t_k b_l, \tag{6.5}$$

where ϵ_{jkl} is the fully asymmetric tensor. As a result of the force, dislocations move on directions dictated by the crystallographic structure. The easiest way to move is to *glide* along the direction of the Burgers vector so that the effective atomic displacements are minimized. It is also possible for a dislocation to *climb* in the orthogonal direction, but this motion entails the displacement of an entire row of atoms. For this reason glide is often the predominant mechanism, while climb occurs because of thermal activation.

As the dislocation line moves across the sample surface it creates a step of one atomic spacing; that is a quantum of plastic strain. The total plastic strain field is then reconstructed by observing the surface morphology and there are simple phenomenological relations that relate the strain rate to the dislocation velocities. According to the Orowan relation

$$\dot{\gamma} = \rho_m b v, \tag{6.6}$$

the plastic strain rate of the material $\dot{\gamma}$ is simply related to average quantities such as ρ_m, the areal density of mobile dislocations, and v, the average velocity along the slip direction. This relation is of mean-field type and does not take into account the fluctuations in the strain, which are very important as we will discuss in the following. Thus it is preferable to replace the average relation by a fluctuating one

$$\dot{\gamma} = \sum_i b_i v_i / A, \tag{6.7}$$

where the sum is over all the dislocations in the slip plane of area A.

In order for a dislocation to move at all, however, the applied stress must overcome the *Peierls* barrier provided by the crystal lattice. This mechanism provides a justification for the presence of a yield stress in semiconductors and BCC metals, but in FCC and HCP metals the actual value of the Peierls stress is much smaller than the observed yield stress. This can be ascribed to the mutual interactions of the disloctions and to the pinning forces due to solute atoms.

A dislocation creates an elastic stress field in the host material, providing an interaction force with the other dislocations through the Peach–Koheler relation in Eq. 6.5. The stress field is obtained solving Eq. 5.7 with the additional constraint

$$\oint d\mathbf{u} = \mathbf{b}, \tag{6.8}$$

where the integral is over a generic contour surrounding the dislocation. In general for a straight dislocation at the origin the stress field in cylindrical coordinates has the form

$$\sigma_{ij} = \frac{Gb f_{ij}(\theta)}{2\pi r}, \tag{6.9}$$

where G is the shear modulus and the angular function f_{ij} depends on the dislocation character. For instance, for a screw dislocation, parallel to the z axis, the only non vanishing components are $f_{z\theta} = f_{\theta z} = 1$. For an edge dislocation with Burgers vector parallel to the x axis, the stress in Cartesian coordinates is given by

$$\sigma_{xy} = \frac{Gb}{2\pi(1-\nu)} \frac{x(x^2 - y^2)}{(x^2 + y^2)^2} \tag{6.10}$$

$$\sigma_{xx} = -\frac{Gb}{2\pi(1-\nu)} \frac{y(3x^2 + y^2)}{(x^2 + y^2)^2} \tag{6.11}$$

$$\sigma_{yy} = \frac{Gb}{2\pi(1-\nu)} \frac{y(x^2 - y^2)}{(x^2 + y^2)^2} \tag{6.12}$$

$$\sigma_{zz} = \nu(\sigma_{xx} + \sigma_{yy}), \tag{6.13}$$

where ν is the Poisson ratio and all the other components are zero. Using these expressions it is possible to compute the resolved shear stresses and hence the interaction forces between the dislocations. Notice that considering infinitely long straight dislocations is an idealization. Although computing the stress field for a generic curved dislocation is a formidable problem, several results are known. We will postpone this discussion to section 6.4 and discuss first the experimental results.

6.3 Acoustic emission and plastic strain bursts in crystals

Plastic deformation is generally viewed as a smooth laminar flow, as opposed to fracture which is associated with rapid and sometimes intermittent crack dynamics. At the atomic scale plasticity corresponds to the individual motion of dislocations and is therefore inhomogeneous. It is thus commonly assumed that above a certain homogenization scale the spatio-temporal fluctuations average out into a fluidlike continuum motion. For a long time, the only well-known exception to this behavior has been the macroscopic plastic instabilities observed in the PLC effect (Zaiser and Hähner, 1997; Ananthakrishna *et al.*, 1999; Ananthakrishna, 2007). This effect consists in repeated stress drops observed in many substitutional metallic alloys in certain ranges of loading and temperature. Depending on these parameters, the stress drops can occur periodically or very irregularly, with sizes that are power-law distributed. The PLC effect is conventionally attributed to the interplay between dislocations and *mobile* solute atoms. When the dislocations are pinned, they attract a solute atom cloud that effectively increases the pinning strength. On the other hand, when the dislocations move, they leave behind the solute cloud, decreasing the pinning strength. This feedback process gives rise to an effective negative strain rate sensitivity, that is believed to be the cause of this stick–slip type of instability. Here we will not explore further the PLC effect, but concentrate instead on the case of standard plasticity where dislocations are the only players.

The general paradigm of smooth plasticity has been challenged by a series of experimental results revealing the presence of strong fluctuations even in stable plastic flow (Weiss, 1997; Weiss *et al.*, 2000; Miguel *et al.*, 2001; Weiss and Marsan, 2003). This has lead us to reconsider the problem from the theoretical point of view, highlighting the collective effect induced by dislocation interactions. While for most practical purposes homogenized engineering modeling techniques are still valid, the role of fluctuations is bound to become a major technological issue in the current trend towards miniaturization. Indeed, experiments on micron-scale samples display constitutive laws that are intrinsically inhomogeneous over all relevant length scales (Uchic *et al.*, 2004; Dimiduk *et al.*, 2006).

Experimentally, the complex character of collective dislocation dynamics can be revealed by (AE) measurements. The acoustic waves recorded in a piezoelectric transducer disclose the pulse-like changes of the local displacements occurring in the material during plastic deformation, whereas a smooth plastic flow would just be detected as a background signal. Thus the method is particularly useful for inspecting possible fluctuations in the dislocation velocities and densities. Ice single crystals and polycrystals have been used as model material to study glide dislocation dynamics by AE (Weiss, 1997; Weiss *et al.*, 2000; Miguel *et al.*, 2001; Weiss and Marsan, 2003) for a series of reasons: Transparency allows direct verification that AE activity is not related to microcracking. There is a range of temperature and stress where diffusion creep is not a significant mechanism of inelastic deformation which, in hexagonal ice single crystals, occurs essentially by dislocation glide on the basal planes along a preferred slip direction. Finally, it is possible to establish an excellent coupling between sample and transducer by fusion/freezing.

Uniaxial compression creep experiments were performed on artificial ice single crystals, employing several steps of constant applied stress. The result is an intense acoustic activity, exhibiting a strong intermittent character (see the inset of Fig. 6.3) (Weiss, 1997; Weiss *et al.*, 2000; Miguel *et al.*, 2001). The energy associated with each acoustic burst displays a power-law distribution spanning several decades

$$P(E) \sim E^{-\tau_E}, \tag{6.14}$$

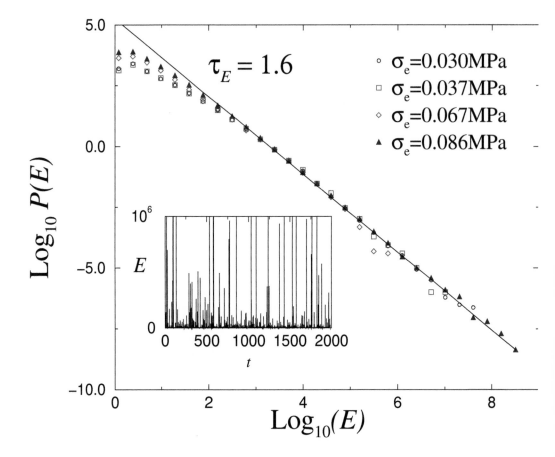

Fig. 6.3 Statistical properties of the acoustic energy bursts recorded in ice single crystals under constant stress. The main figure shows the distribution of energy bursts for the different loading steps. While the cutoffs depend slightly on the applied stress, the power-law exponent remains the same. The fit yields an exponent $\tau_E = 1.60 \pm 0.05$. In the inset we report a typical recorded acoustic signal. Source: Miguel *et al.* (2001)

with $\tau_E = 1.6$ as shown in Fig. 6.3. Similar measurements were also performed for polycrystalline ice and the result is analogous but the exponent is reduced (i.e. $\tau_E = 1.35$) in comparison with single crystals (Richeton *et al.*, 2005). In addition, it was possible to test the effect of grain size on the AE distributions, and it was observed that the cutoff to the power-law behavior increases with the grain size (Richeton *et al.*, 2005). This may indicate that the grain boundaries act as strong pinning centers for the dislocations, impeding their motion.

The fluctuating character of plastic flow can also be inferred from other experimental methods beside AE. The history of plastic deformation is encoded in the sample surface, where moving dislocations leave atomic scale steps (slip lines). Surface steps can be analyzed with great detail using a variety of techniques, from old-fashioned surface micrographs to transmission electron microscopy, atomic force microscopy, and scanning white-light interferometry. The experiments typically show incredibly complex slip patterns that can be characterized by scaling laws. Under plastic deformation the surface of a sample roughens, becoming self-affine over a wide range of scales (Zaiser *et al.*, 2004). In addition, the distribution of surface steps decays as a power law, that should be related to dislocation avalanches (Schwerdtfeger *et al.*, 2007). With a higher spatio-temporal resolution it will soon become possible to visualize the dislocation avalanches, by imaging the surface profiles at different times.

Important progress to reveal dislocation avalanches has been accomplished for plastically deformed micron-sized samples (see Fig. 6.4) (Uchic *et al.*, 2004; Dimiduk *et al.*, 2006) where the stress–strain curves have a staircase character. A step $\Delta\gamma$ in the stress–strain curve can be directly related, using Orowan relation (Eq. 6.7), to the dislocation avalanche size

$$s \equiv \Delta\gamma = \int^T dt\dot{\gamma} = \int^T dt \sum_i b_i v_i(t)/A, \qquad (6.15)$$

where T is the avalanche duration and A is the area of the slip plane. The avalanche size distribution determined from these experiments is a power law with an exponent $\tau \simeq 1.53$ (Dimiduk *et al.*, 2006) (see Fig. 6.4). The reduced sample size is a key factor to observe these fluctuations, that are too small to be perceived in standard mechanical tests on larger samples, where the stress–strain curve looks smooth. Fluctuations are revealed, however, by AE measurements even in macroscopic samples. The reason is the following: The dissipated energy E during an avalanche is proportional to the strain step s: $E = \sigma s V$, where V is the sample volume and σ is the applied stress. Hence, the cutoff of the avalanche energy distribution should increase with the volume as $E_0 \propto s_0 V$, where s_0 is the cutoff in the strain avalanche distribution, that in fact decreases with the sample size as $1/L$ (Csikor *et al.*, 2007). If we assume that the acoustic energy is proportional to the dissipated energy, we understand why acoustic avalanches are observed in macroscopic single crystals, while strain avalanches are not. Notice that in several instances, papers in the literature report the distribution of AE amplitudes $P(a)$. Amplitudes and energies are related by $E = a^2$. It is thus possible to derive a scaling relation for the exponent τ_a of the amplitude distribution. Conservation of probability imposes that $P(a)da = P(e)de$. Using the relation between e and a we then obtain $\tau_a = 2\tau - 1$. Hence, $\tau = 3/2$ corresponds to $\tau_a = 2$, in agreement with experiments (Weiss, 1997). Notice also that since energy is proportional to strain, we must have $\tau_E = \tau$ as suggested by the experiments.

Small samples are very interesting because the effect of fluctuations is enhanced and in addition to a staircase in stress–strain curve one observes strong sample-to-sample fluctuations and size effects: smaller samples are stronger. This effect is also observed in fracture, where it is usually associated with extreme value statistics, as discussed in Chapter 5. The origin of this phenomenon in plasticity is not fully understood but is of primary importance in the current trend towards nanoscales.

6.4 Depinning of dislocation lines and assemblies

Dislocations are deformable linear defects interacting with impurities in the material and it is thus possible to interpret their behavior in the general framework of the

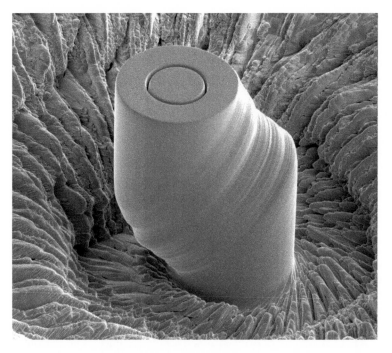

Fig. 6.4 A plastically deformed Ni sample with diameter $d = 20\mu m$. Notice the steps on the surface induced by dislocation motion. Source: Image courtesy of Michael Uchic, Air Force Research Laboratory, Materials and Manufacturing Directorate. See also Dimiduk *et al.* (2006).

depinning transition. Following this idea, the scaling observed in dislocation avalanches can be related to the universality classes discussed in Chapter 4. In the metallurgical literature the problem of dislocation depinning is known as *solid solution hardening*, since the presence of solute atoms in the crystal provides pinning centers which increase the yield stress of the material, making it harder.

6.4.1 Dislocation pinning mechanisms

The presence of solute atoms changes the local properties of the host material, resulting in a force on nearby dislocations. A simple model to quantify these interactions assumes that the impurity can be represented by a sphere placed in the host material, treated as an elastic continuum. Normally there is a size misfit between the atoms of the host material and the solute atoms, represented by an effective volume change ΔV. An edge dislocation induces an elastic stress in its surroundings whose hydrostatic component is given by

$$p = \frac{1+\nu}{1-\nu} \frac{Gb}{2\pi} \frac{\sin\theta}{r},$$
(6.16)

where (r, θ) is the position of the solute atom with respect to the dislocation. The work associated with this stress is given by $W = p\Delta V$, which can then be used to compute the effective interaction energy or the force between the dislocation and the impurity.

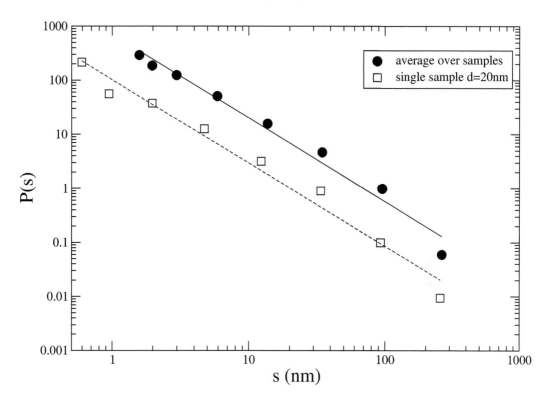

Fig. 6.5 The distribution of avalanche sizes measured in a single sample of diameter $d = 20\mu m$ and an average over different samples. A fit of the power-law decay yields $\tau = 1.53$. Source: Data from Dimiduk *et al.* (2006).

For instance, an edge dislocation lying in the x direction with the Burgers vector in the y direction with a glide plane at a distance z from the solute atom is subject to a pinning force given by

$$F_p = \frac{Gb\Delta V}{\pi z^2} g(y/z), \qquad (6.17)$$

where $g(u) \equiv 2u/(1 + u^2)$. Thus if ΔV is negative the dislocation is attracted toward the impurity and is repelled otherwise. Assuming that ΔV and z are of the order of the lattice constants (i.e. b) one can estimate that the effective pinning strength is of the order of $F_p \simeq Gb^2\delta$, where δ is the relative volume change due to the solute atom.

The presence of a solute atom does not only involve a change in the effective volume, but also a local modification of the elastic moduli. Since the elastic energy density associated with the dislocation depends linearly on the shear modulus G, a local variation of G induces an effective interaction. For instance, the dislocation would be attracted to regions of lower G and repelled by regions of larger G. In order to obtain a quantitative estimate of the pinning force one can consider an effective medium description, yielding a result similar to the one reported in Eq. 6.17. The pinning strength in this case is of the order of $F_p \simeq Gb^2\delta$, where δ is a measure of the relative modulus change.

In ionic solids, an important source of pinning comes from the electrostatic inter-action between a dislocation and charged impurities. In particular, an edge dislocation creates a charge modulation which gives rise to an effective interaction force with the solute ions. Electrostatic interactions play a minor role in metals, due to differences in the atomic valence between the solute and host atoms.

Above we have discussed only the pinning forces due to solute atoms, but one should bear in mind that other sources of pinning may be present in the system. In two-phase alloys with a small volume fraction of one of the constituents, particles of the second phase act as pinning centers for the dislocations. The pinning mechanism is similar in principle to that of solute atoms, although the size of pinning centers is larger. In particular, pinning forces arise due to size or elastic moduli misfit, differences in the atomic structure between the two phases, and other effects. Depending on the properties of the two, phases dislocations can either permeate the particles or not. In the second case, in order to overcome the obstacle, the dislocation leaves an Orowan loop encircling the particle.

Immobile dislocations threading the glide plane of a mobile dislocations are usually referred to as a forest and can be treated as a set of pinning centers. The pinning force is due to several mechanisms. First, one should consider the Peach–Koheler force due to the long-range elastic stress field induced by the forest. To estimate this term one should bear in mind that dislocations will not be found at uncorrelated random positions, since this configuration would produce a diverging self-stress. In practice, dislocations arrange into screened configurations, such as dipoles or walls. A simple way to take into account this observation was proposed by Wilkens and amounts to placing dislocations randomly into boxes of fixed size (Wilkens, 1976). This yields a random stress field with zero mean, short range correlations, and a variance

$$\langle \sigma_p^2 \rangle = CbG\rho \log(\xi/b), \tag{6.18}$$

where ρ is the dislocation density, ξ is the correlation length, and C is a numerical constant. In addition to this effect, when the moving dislocation comes close to a dislo-cation in the forest, microscopic processes such as jog or junction formation dominate the pinning strength. Quantitative estimates of forest pinning should take into account the character and orientation of the dislocations. One should bear in mind, however, that the distinction between mobile and immobile dislocations is quite arbitrary, since initially immobile dislocation can in principle set into motion.

Finally, the atomistic structure of the crystal is responsible for the Peierls–Nabarro force , which is particularly important in BCC crystals but can be neglected in FCC crystals. As the dislocation glides it has to overcome an energy barrier separating one lattice plane to the next. This effect can be taken into account by a periodic force oscillating with the lattice planes, whose amplitude can be estimated and depends on the particular lattice structure. Typically one finds that this amplitude decreases exponentially with the ratio between the crystal spacing in the direction perpendicu-lar and parallel to the glide direction. For instance, the Peierls–Nabarro on an edge dislocation could be approximated as

$$F_{PN}(y) = \frac{Gb^2}{2\pi} \exp(-2\pi ya/b) \sin(2\pi y/b), \tag{6.19}$$

where a is the lattice spacing in the direction normal to the glide plane. Differently from solute atom interactions, the Peierls–Nabarro force is not random, but can nonetheless give rise to pinning.

6.4.2   Self-stress and line tension

A dislocation interacting with a pinning center, such as a solute atom, a particle, or a forest dislocation, typically does not remain straight. Bending a dislocation has an energy cost, which can be used to balance the pinning energy. In the simplest picture, the dislocation can be treated as an elastic string with a line tension Γ. In this case, the energy cost for increasing the dislocation length by an amount dl is given by $dE = \Gamma dl$. The problem of estimating the line tension Γ from linear elasticity theory has been addressed in the past using several approaches. From purely dimensional arguments, we would expect that $\Gamma \propto Gb^2$, but quantitative calculation shows that Γ depends on the particular deformation mode of the dislocation. Thus the line tension model can only be considered as an approximation for the long-range self-stress of the dislocation.

Expressions for the self-stress of a dislocation were reported by DeWit and Koehler (1959), Brown (1964), Brailsford (1965), and Foreman (1967). Consider a dislocation lying in the xy plane, with the Burgers vector forming an angle α with the x direction, and denote by $(x(s), y(s), 0)$ its position as a function of the line coordinate s (see Fig. 6.6). In this case the self-stress of the dislocation is given by (Foreman, 1967)

$$\sigma_{xz}(x_0, y_0) = \frac{Gb}{4\pi(1-\nu)} \int_{line} ds \left[\left(\frac{dy}{ds} \right) \frac{x_0 - x(s)}{(r_0 - r(s))^3} - \left(\frac{dx}{ds} \right) \frac{y_0 - y(s)}{(r_0 - r(s))^3} \right] \times$$

$$\left[1 + \nu - 3\nu \left(\frac{y_0 - y(s)}{r_0 - r(s)} \cos\alpha - \frac{x_0 - x(s)}{r_0 - r(s)} \sin\alpha \right) \right]. \tag{6.20}$$

From Eq. 6.20, one can compute the elastic self-energy and realize that the result is not in the form of $\Gamma \int_{line} ds$ as in the line tension approximation.

To fix the ideas, consider an edge dislocation in the xy plane with Burgers along y and impose small deformations around the configuration in which the dislocation is aligned along x at $y = 0$. For simplicity, we do not consider overhangs, so that the dislocation can be parametrized by a single value function $y = u(x)$. The y component of the restoring force on the dislocation is obtained from Eq. 6.20 expanding to lowest order in the deformation,

$$f_y(x) = \frac{Gb^2}{4\pi} \int dx' \left[\left(\frac{du}{dx} \right) \frac{x - x'}{(1-\nu)|x' - x|^3} - \frac{u(x) - u(x')}{(x - x')^3} \right]. \tag{6.21}$$

It will be convenient to express the force in Fourier space, and expand to lowest order in q

$$\tilde{f}_y(q) = \frac{Gb^2(2\nu - 1)}{4\pi(1-\nu)} q^2 \log(|q|a) \tilde{u}(q), \tag{6.22}$$

where a is a low scale cutoff, of the order of b.

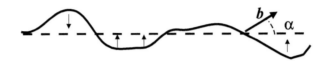

Fig. 6.6 A configuration of a deformed dislocation. The burgers vector **b** of the dislocation and the axis of the undeformed dislocation (dashed line) define an angle α. The small arrows indicate the local restoring force.

In the line tension approximation, the relevant component of the elastic force is now given to lowest order in Fourier space by

$$\tilde{f}_y^{lt}(q) = -\Gamma q^2 \tilde{u}(q). \tag{6.23}$$

Comparing Eqs. 6.22 and 6.23 we see that although the self-stress can not be directly mapped into a line tension, a good approximation can be obtained taking

$$\Gamma = \frac{Gb^2(2\nu - 1)}{4\pi(1 - \nu)} \log(L/a), \tag{6.24}$$

where L is the upper cutoff or the relevant length scale of the problem. Clearly, different approximations should be used for dislocations of different orientation and shape than the one considered here.

As shown above, in the case of an isolated dislocation the line tension approximation is correct up to logarithmic factors, which in most cases will not play a major role. The situation is different for a dislocation assembly, where the long-range character of the elastic stress field can give rise to important corrections to the self-stress. An example is provided by a dislocation wall, or a low-angle grain boundary. For simplicity let us consider a set of edge dislocations with Burgers vector in the y direction, lying on equally spaced xy planes. The distance between the glide planes of the dislocations along the z axis is taken to be D (see Fig. 6.7).

Representing the dislocation displacements by a function $u_m(x)$, with m indicating the different planes, one finds to linear order in u that the self-stress is given by

$$\sigma_{xz}(x, u_n(x), nD) = \frac{Gb}{4\pi(1 - \nu)} \sum_m \int dx' \frac{x - x'}{R_{nm}^3} \left[1 - 3\frac{D^2(n - m)^2}{R_{nm}^2} \right] \times$$

$$\left[\partial_x' u_m(x') - (1 - \nu) \frac{u_n(x) - u_m(x')}{R_{nm}^3} \right], \tag{6.25}$$

where $R_{nm} \equiv \sqrt{(x - x')^2 + D^2(n - m)^2}$. If we proceed as in the case of an isolated dislocation and compute the force in Fourier space, the leading contribution at large wavelengths does not scale as $q^2 \log q$ but is given by

$$\tilde{f}_y(q_x, q_z) = \frac{Gb^2}{4D(1 - \nu)|q|}(q_z^2 + (1 - \nu)q_x^2)\tilde{u}(q_x, q_z). \tag{6.26}$$

The factor $|q|$ at the denominator makes it impossible to define an approximate wall surface tension for the low-angle grain boundary, as for an isolated dislocation, and it is necessary to deal directly with the non-local long-range self-stress.

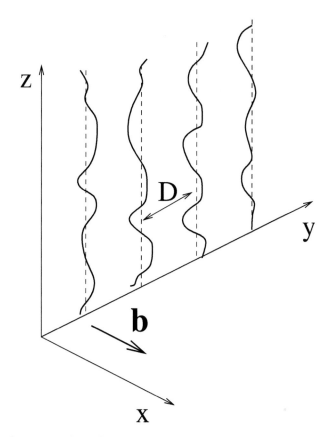

Fig. 6.7 A configuration of a deformed low-angle grain boundary. Each dislocation deforms along the y axis, coinciding with the direction of the Burgers vector $\vec{b}$. A dislocation pileup is obtained when b and the dislocation displacement vectorare along the x direction. Source: Moretti *et al.* (2004).

The same situation occurs in a dislocation pileup, where a large number of dislocations with the same Burgers vector are piled up in their glide plane. In the approximation of an infinite, equally spaced, pileup the calculation of the self-stress follows the one for a low-angle grain boundary, and again the self-stress kernel scales as $|q|$. Notice that a similar results hold for pileups and walls made of rigid dislocations, that can hence be treated as points moving in one and two dimensions respectively. The self-stress can be simply computed summing up all the stresses due to each dislocation, and the Fourier transform is given by

$$\tilde{f}_y(q) = \frac{Gb^2}{4D(1-\nu)}|q|\tilde{u}(q), \tag{6.27}$$

and again the line tension approximation makes no sense. Finally, we notice that more complicated dislocation structures, with dislocation of different orientations and

Burgers vectors, will lead to more complicated forms of the self-stress, but at present this formidable problem has not yet been addressed.

6.4.3 Dynamics of dislocations in the presence of impurities

The dynamics of individual dislocations and dislocation arrays in presence of solute atoms falls in the realm of general depinning systems, and we can thus directly apply the theoretical machinery developed in Chapter 4. If we assume a linear force-velocity law, dislocation glide follows

$$\chi \frac{\partial u(x,t)}{\partial t} = b[\sigma_e + \sigma_{int}(u) + \sigma_{pin}(x,u)], \tag{6.28}$$

where χ is a damping constant, σ_e is the externally applied resolved shear stress, σ_{int} is a non-local interaction kernel obtained from Eq. 6.20, and σ_{pin} is the pinning stress discussed in section 6.4.1. Owing to the short-range character of the interaction kernel, we expect that Eq. 6.28 is in the same universality class as a linear interface with local interactions and we can thus read off the exponent from Table 4.3. Here we mention that avalanches should be power-law distributed with exponents $\tau = 1.05$ and $\alpha = 1.0$ for sizes and durations, respectively. The cutoff sizes are ruled by the externally applied stress, defining the exponents σ and Δ. In the case of dislocation arrays the situation remains qualitatively the same, but due to the long-range character of the interaction kernel, scaling as $|q|$ in Fourier space, the exponents are described by mean-field theory, up to logarithmic corrections. Hence, from Table 4.3 we read off $\tau = 3/2$ and $\alpha = 2$ for the avalanche exponents.

An interesting variation of this problem is found when the solute atoms are mobile and perform a thermally assisted diffusive motion biased by the force field due to the dislocations. This general problem has a long history due to its relevance for the Portevin–Le Chatelier effect. Using simple considerations it is widely accepted that the presence of diffusing solute atoms leads to strain-rate sensitivity and consequently to plastic instabilities, as mentioned in section 6.1. The reason for this behavior can be understood intuitively: A pinned dislocation will accumulate a solute cloud around it and become even more difficult to unpin. On the other hand, if the solute atom diffusivity is low, a moving dislocation interacts with fewer impurities, so that the effective *dynamic* pinning force is smaller than the static one. The net result is a flow stress that decreases with strain rate. There has been a renewed effort to import this general understanding into interacting dislocation models and study the statistical properties of the strain fluctuations.

6.5 The yielding transition and dislocation avalanches

The description of a single dislocation, or of an ordered dislocation array, moving in glide plane interacting with a random stress field provided by solute atoms and immobile dislocations, is theoretically appealing, but represents a strong idealization. In most cases we must deal with a disordered assembly of interacting dislocations moving collectively in different glide planes. This general problem is difficult to address even numerically, although much progress has been done in simulating large collections of interacting dislocation lines in three dimensions. Here, we consider the conceptually simpler case of a two-dimensional system in which a set of dislocations are moving along their glide direction. Next, we discuss how the observed *jamming* behavior can be interpreted in the framework of continuum plasticity as a general *yielding transition*.

6.5.1 Creep relaxation of dislocation systems

In order to reproduce the avalanche behavior observed in plasticity, one can perform discrete dislocation dynamics (DDD) simulations, in which one follows the dynamics of an ensemble of interacting dislocations. The simplest DDD model can be thought to represent the cross section of a single-slip oriented crystal where N point-like edge Dislocations glide in the xy plane along directions parallel to the x axis. dislocations with positive and negative Burgers vectors $\mathbf{b}_n = \pm b\hat{x}$ are assumed to be present in equal numbers, and the initial number of dislocations is the same in every realization. An edge dislocation with Burgers vector $b\hat{x}$ located at the origin gives rise to a force at a point $\mathbf{r} = (x, y)$ given by Eq. (6.5). One further assumes an overdamped dynamics in which the dislocation velocities are linearly proportional to the local forces. Accordingly, the velocity of the nth dislocation along the glide direction, if an external shear stress σ_e is also applied, is given by

$$\chi v_n = b_n [\sigma_e + \sum_{m \neq n} \sigma_{xy}(\mathbf{r}_{nm})], \qquad (6.29)$$

where χ is the damping constant, σ_{xy} is the relevant component of the shear stress (see Eq. 6.13) and $\mathbf{r}_{nm} \equiv \vec{r}_n - \vec{r}_m$ is the relative position vector of dislocations n and m. Periodic boundary conditions are imposed in the direction of motion (i.e. the x axis). In order to take correctly into account the long range nature of the elastic interactions, the stress should be summed over an infinite number of images. When the distance between two dislocations is of the order of a few Burgers vectors, linear elasticity theory (i.e. Eq .(6.5)) breaks down. In these instances, phenomenological nonlinear reactions, such as the annihilation of a pair of dislocations, describe more accurately the real behavior of dislocations in a crystal. In this model, one *annihilates* a pair of dislocations with opposite Burgers vectors when the distance between them is shorter than a cutoff y_e Miguel *et al.*, (2001, 2002). In addition, one includes in the model dislocation multiplication introducing a pair-creation rate r.

Miguel *et al.* (2002) report the temporal relaxation of a simple dislocation dynamics model discussed above. When multiplication is absent (i.e. $r = 0$), the strain rate (defined as in Eq. 6.7) initially decays as a power law with an exponent close to $2/3$, recovering the experimentally observed Andrade creep law. At larger times, the strain rate is observed to cross over to a linear creep regime (i.e. to a plateau signaling a steady rate of plastic deformation) whenever the applied stress is larger than a critical threshold σ_c, or, otherwise, to decay exponentially to zero (see Fig. 6.8).

These results suggest that a possible interpretation of dislocation motion and the corresponding creep laws of crystalline materials could also be found within the general *jamming* framework proposed to encompass a wide variety of non-equilibrium soft and glassy materials (Liu and Nagel, 1998). Most of these physical systems consist of various types of soft particles closely packed into an amorphous state. At such high concentrations, the relative motion of these particles is drastically constrained and,

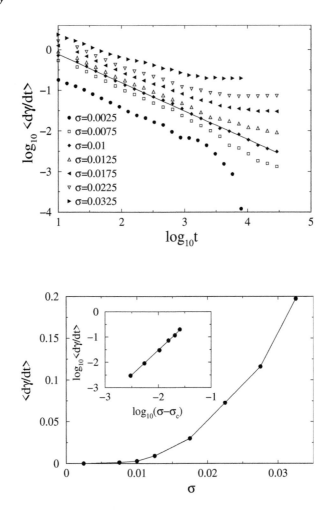

Fig. 6.8 (top) The strain-rate relaxation for different stress values. The system size is $L = 100b$ and the initial density of edge dislocations is around 1%. The solid line is the Andrade law $d\gamma/dt \sim t^{-2/3}$. (bottom) The steady state value of the strain rate as a function of the applied stess. The inset shows the behavior as a function of $\sigma - \sigma_c$ in log-log scale. Source: Miguel *et al.* (2002).

as a consequence, soft and concentrated materials usually respond like elastic solids upon the application of low stresses. One then says that the system is jammed, since it is unable to explore all the available configuration space. On the other hand, they flow like viscous fluids above the so-called yield stress value σ_y, exhibiting a common rheology. Light-scattering experiments (Kegel and van Blaaderen, 2000; Weeks *et al.*, 2000) allow, for instance, the detection of the intermittent dynamics of a gel formed from attractive colloids, suggesting that intermittent behavior seems to be a fundamental ingredient for the slow relaxation of jammed materials. Moreover, this unjamming transition can be induced by changing either the external stress applied, the density, or the temperature of the system. The analogies of dislocation motion and these so-called jammed systems are further explored by considering the influences of dislocation multiplication, and thermal-like fluctuations on the dynamics. Dislocation multiplication favors the rearrangements of the system and induces a linear creep regime (flowing phase) at lower stress values, but it does not affect the initial power-law creep. The introduction of a finite effective temperature T has a similar effect (Miguel *et al.*, 2002).

6.5.2 Strain bursts in dislocation systems

While the average creep relaxation of the dislocation dynamics model is well described by the smooth Andrade law, the velocity signal is in general strongly fluctuating. Due to their complicated mutual interactions, dislocations are usually jammed into metastable configurations, formed by walls, dipoles, and far more complex dislocation structures. The formation of metastable configurations is responsible for the quiescent intervals in the acoustic activity, followed by bursty events that occur when the accumulated shear stress and/or the multiplication and annihilation processes can eventually favor the partial destruction and/or rearrangement of the dislocation structures. Thus the complex features of AE can be explained by the jamming of dislocations and their subsequent slip, giving rise to avalanche-like events.

A detailed study of the avalanche statistics associated with two-dimensional dislocation systems was performed in Ispánovity *et al.* (2014) where direct integration of Eq. 6.29 was complemented by simulation of cellular automaton models defined by discretizing the system in space and time. In the cellular automaton, dislocations are allowed to move from one cell to a neighboring cell if such a move decreases the elastic energy of the system. Two different rules are applied for the dynamics: First, in extremal dynamics at each step only the move which produces the largest energy decrease is carried out. Second, in random dynamics, the moved dislocation is selected randomly from those that are allowed to move. A quasistatic stress-controlled loading protocol was implemented both for the continuous time model and for the cellular automata. In short, a random initial dislocation configuration is allowed to relax at $\sigma_e = 0$ into a metastable arrangement. Then σ_e is increased until the most unstable dislocation starts to move. At that point the external stress is kept constant until the avalanche stops and system enters again into a jammed state. After the avalanche is terminated, the external stress is again increased at a slow rate until the next avalanche is triggered (Ispánovity *et al.*, 2014).

For each variant of the model, the avalanche size distribution $P(s)$ was computed at different levels of external stress below the yield stress, and was shown to be well characterized by a power law with a cutoff,

$$P(s) \propto s^{-\tau} f(s/s_0). \tag{6.30}$$

The exponent τ and the cutoff values s_0 were then obtained at different stress levels and system sizes. The cutoff s_0 was found to follow a size-dependent scaling law

$$s_0(\sigma_e, N) \propto N^{\beta} \exp(\sigma_e/\sigma_0), \tag{6.31}$$

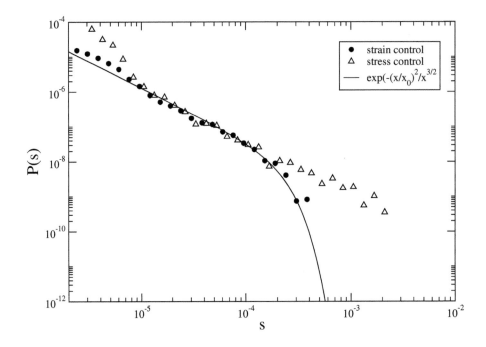

Fig. 6.9 The distribution of avalanche sizes for for a three-dimensional model of interacting dislocation lines under strain control and stress control. Both distributions scale with an expoenent $\tau = 1.5$, but stress-control data have a larger cutoff. Source: Csikor *et al.* 2007

where N is the number of dislocations. The power-law exponent was estimated to be $\tau \approx 1.0$, clearly different from the mean-field theory prediction ($\tau = 1.5$). When the avalanche distribution is obtained by integrating over all the values of the external stress the exponent is larger $\tau_{int} \approx 1.3$, as expected. Moreover, the avalanche size was found to scale with the duration as $s \propto T^{\gamma_{st}}$, with $\gamma_{st} \approx 1.35$. These results imply that there are fundamental differences between the behavior of dislocation systems and the interface pinning/depinning scenario. Of particular interest is the behavior of the cutoff of the avalanche size distribution which, rather than diverging at some critical stress σ_c as for depinning interfaces, scales exponentially with stress but diverges with system size at every stress level (Ispánovity *et al.*, 2014).

Plastic avalanches have also been studied in three-dimensional dislocation dynamics simulations (Csikor *et al.*, 2007; Lehtinen *et al.*, 2016). These kind of models are extremely challenging from the computational point of view and for this reason one has access only to relatively small sample sizes. On the other hand, they provide a more faithful representation of the dynamics of interacting dislocations, including the formation of junctions, the multiplication of dislocation through Franck–Read sources, and the effect of the boundary conditions on the elastic stresses. Early simulations showed that the scaling features of the avalanche distribution persist in three dimensions, with an exponent $\tau \simeq 1.5$ (Csikor *et al.*, 2007). In addition, the exponent was found to be the same for single slip and multiple slip, and independent of the presence of cross-slip and of the loading mode. On the other hand, the value of the cutoff was found to change with the loading mode (see Fig. 6.9). The robustness of the exponent value is supported by the agreement between experiments and model. Guided by numerical simulation results, it was possible to obtain the scaling of the cutoff of the avalanche distribution, given by

$$s_0 = \frac{bE}{L(\Theta + \Gamma)}, \tag{6.32}$$

where E is the Young modulus, L is the linear size of the system, Θ is the hardening coefficient, and Γ is the stiffness of the traction machine when the sample is loaded under strain control (Csikor *et al.*, 2007). Using this expression it is possible to collapse all the simulated and experimental distribution into a single master curve (see Fig. 6.10). This general scaling law was later confirmed by other independent three-dimensional dislocation dynamics simulations (Devincre *et al.*, 2008) and by experiments in Mo and Au micropillars (Brinckmann *et al.*, 2008).

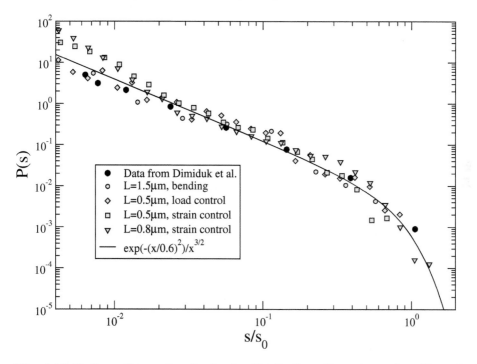

Fig. 6.10 Scaling collapse of avalanche size distributions. Open data points: Data obtained from simulations of systems of different sizes in load and displacement control. Scaling parameters: $b = 2.8 \times 10^{-10}$ m (Al); $\Gamma = E$, $\Theta = E/10$ (displacement-controlled tension/compression and bending); $\Gamma = 0$, $\Theta = E/10$ (load-controlled tension/compression). Full data points: Experimental data of Dimiduk et. al. (Dimiduk *et al.*, 2006), scaling parameters: $b = 2.5 \times 10^{-10}$ m (Ni), $\Gamma = 0$, $\Theta = E/1000$ (load controlled compression). Full line: Scaling function $P(S) \propto S^{-1.5} \exp(-(S/0.6)^2)$. Source: data from Csikor *et al.* (2007).

These findings have important implications for deformation processes on the micron scale. As a consequence of the stochastic and intermittent nature of the deformation process, the deformation behavior of a small enough sample can no longer be predicted in a deterministic sense. As the maximum avalanche strain increases with decreasing system size, the stochastic heterogeneity of deformation becomes more and more pronounced. Hence it may be difficult, on the micron and sub-micron scale, to control the results of plastic-forming processes. This problem could be relevant in the processing of micron-scale components such as bonding wires in chips.

The universality class of three-dimensional dislocation systems was later reconsidered in the light of the unconventional scaling observed in two dimensions (Ispánovity *et al.*, 2014). Contrary to previous three dimensional simulations (Csikor *et al.*, 2007), new simulations were performed with periodic boundary conditions (Lehtinen *et al.*, 2016). In this case, it was shown that while the avalanche distribution exponent was apparently close to the mean-field value $\tau = 3/2$, the results were better interpreted as a crossover between a higher value $\tau \simeq= 1.8$ at small sizes and a smaller value at larger sizes $\tau = 1.1$ (Lehtinen *et al.*, 2016). Furthermore, in analogy with two-dimensional results (Ispánovity *et al.*, 2014) it was observed that the typical avalanche size was increasing roughly exponentially with the external stress for all system sizes (quantified by the number of dislocations N present in the system) and at any given stress it was found to be significantly dependent on N (Lehtinen *et al.*, 2016). Notice, however, that in contrast with previous simulations where the avalanche signal was stationary in time - the new simulations clearly displayed a non-stationary signal. All the differences are probably due to the different loading and boundary conditions employed in the two studies.

All the models used to study dislocation avalanches assume that the main mode of plastic deformation is associated with fast dislocation glide processes. Concurrently and between these fast glide events, other background processes contribute to plastic flow. We should consider for instance thermally activated processes such as the viscoelastic response of the dislocation forests, the localized dislocation climb motion in directions different from the main glide plane, or cross-slip events of dislocations shifting them between glide planes. All these processes contribute to a non-trivial strain rate and time dependence of the dislocation avalanches (Papanikolaou *et al.*, 2012). Experiments in compressed microcrytals and simulations show unconventional quasi-periodic avalanche bursts, associated with avalanche distribution exponents that *increase* as the strain rate is increased, in contrast with a general argument discussed in this book (see section 1.6) postulating that exponents should decrease with the driving rate (White and Dahmen, 2003). This peculiar behavior was named the *self-organized avalanche oscillator* and was shown to represent a general class of non-equilibrium critical states exhibiting oscillatory approaches towards a critical point (Papanikolaou *et al.*, 2012). Beside crystal plasticity, similar processes could be relevant to explain the occurrence quasi-periodic earthquakes. The interested reader can find detailed discussions of plasticity from a statistical physics point of view also in Zaiser (2006) and Sethna *et al.* (2017).

6.5.3 The yielding transition in continuum plasticity

The analysis reported in the previous section shows that there are intriguing analogies between the glide motion of a deformable dislocation line in the presence of disorder and the dynamics of an assembly of interacting dislocation moving on different glide planes. This analogy has been reformulated into a more rigorous framework by Zaiser and Moretti (2005), employing the formalism of continuum plasticity discussed in Sec 6.1. They consider a single-slip system (along the x direction) and describe the plastic strain by a field $\gamma((r))$, assuming it independent of the z coordinate. These conditions closely represent the geometry of the dislocation dynamics model discussed above. Employing a general viscoplastic constitutive law, one can write

$$\frac{\partial \gamma}{\partial t} = C[\sigma_e + \sigma_{int}(r, \gamma) + \sigma_p(r, \gamma)], \tag{6.33}$$

where σ_e is the externally applied stress and σ_p is the effective pinning stress due to immobile threading dislocations, whose fluctuations are quantified in Eq. 6.18, or by solute atoms. The internal stress induced by the local variations in the strain field $\gamma(\mathbf{r})$ can be computed in the framework of linear elasticity and for the single slip system considered here it is given by (Zaiser and Moretti, 2005)

$$\sigma_{int}(\mathbf{r}, \gamma) = \frac{G}{2\pi(1-\nu)} \int d\mathbf{r}' \left[\frac{1}{(\mathbf{r} - \mathbf{r}')^2} - \frac{8(x - x')^2 (y - y')^2}{(\mathbf{r} - \mathbf{r}')^6} \right] \gamma(\mathbf{r}')$$

$$+ \frac{G}{(1-\nu)}[\langle \gamma \rangle - \gamma(\mathbf{r})], \tag{6.34}$$

where $\langle \gamma \rangle$ is the volume average of the plastic strain field. Notice that the second term is of mean-field type, while the first term is long-range and anisotropic. In Fourier space the internal stress scales as $\tilde{\sigma}_{int} \propto \tilde{\gamma}_{\mathbf{k}} k_x^2 k_y^2 / |k|^4$ and thus has effectively $a = 0$ in the notation of Chapter 4. Eq. 6.33 is thus equivalent to the one describing an elastic manifold with non-local stiffness moving in a disordered landscape under the action of an external force. Owing to the long-range interactions we may expect to recover mean-field critical behavior, and simulations display a size distribution decaying as a power-law distribution with exponent $\tau \simeq 3/2$ (Zaiser and Moretti, 2005). Yet, in the model the long-range interaction kernel is not convex as in standard depinning models. This induces deviations from mean-field theory that we will discuss in the following section in the context of amorphous plasticity.

It is important to emphasize here the crucial role played by work hardening in the yielding transition. As discussed in section 6.1, in most materials dislocation proliferation effectively increases the stress necessary to sustain dislocation motion. This effect is usually modeled by increasing the flow stress or equivalent by introducing in Eq. 6.34 an additional back-stress $\sigma_B = -\Theta\gamma$, where Θ is the hardening coefficient. Thus if we ramp up the external stress the back-stress keeps the effective stress $\sigma_e + \sigma_B$ close to the critical yield stress. We easily recognize here the typical SOC mechanism already discussed in detail for sandpile models, interfaces, and so on. We could thus expect that the avalanche cutoff is controlled by Θ, playing the role of dissipation in sandpile models. This is also confirmed by the fact that the cutoff scales with $1/\Theta$ (see Eq. 6.32

when $\Gamma = 0$) (Zaiser, 2006; Csikor *et al.*, 2007). In order to refine the description of the yielding transition, one can consider also plastic strain gradients which arise due to small-scale interactions, not captured by linear elasticity. In the simple approximation one can add to Eq. 6.34 a term proportional to $\partial^2\gamma/\partial x^2$ which, although irrelevant for the scaling behavior, is important to account for the slip patterns observed in experiments. While we can expect that the yielding transition represents a useful theoretical framework to understand plastic flow in general, the detailed implementation of the program for a generic multiple-slip system of interacting curved dislocation appears to lie well beyond present possibilities.

6.6   Strain bursts in amorphous plasticity

Amorphous materials, such as silica glasses or bulk metallic glasses, display high strength and hardness, and remarkable corrosion and wear resistance (Trexler and Thadhani, 2010). When deformed beyond the elastic limit, amorphous solids typically fail catastrophically in a brittle fashion due to the formation and propagation of localized shear bands (Schuh *et al.*, 2007; Chen, 2008). Under particular experimental conditions of strain rate and temperature, however, amorphous materials may deform plastically, but due to the absence of a crystal structure it is not possible to associate plastic deformation with the motion of dislocations. Plastic strain bursts are observed in the compression of amorphous nanopillars and in bulk metallic glasses (Shan *et al.*, 2008; Wang *et al.*, 2009; Sun *et al.*, 2010) in analogy with similar observations made in crystalline materials (Dimiduk *et al.*, 2006). Experimental measurement of the deformation of bulk metallic glasses obtained at high temporal resolution displayed power law distributed bursts with exponents compatible with mean-field scaling (Antonaglia *et al.*, 2014*a*, 2014*b*), although the relatively small scaling range could lead to different interpretations regarding the universality class.

Atomistic simulations of plastically deformed amorphous materials displayed intermittent strain bursts similar to those observed in experiments, but the universality class of the associated scaling behavior is still hard to pinpoint (Maloney and Lemaitre, 2004; Demkowicz and Argon, 2005; Maloney and Lemaître, 2006; Bailey *et al.*, 2007; Maloney and Robbins, 2009). Large-scale three-dimensional simulations of simple model glasses under shear (e.g. binary mixtures with simple attractive pair interactions) show that the damping constant changes the scaling exponents (Salerno *et al.*, 2012; Salerno and Robbins, 2013). Furthermore, the results obtained in the overdamped regime show significant deviations from mean-field scaling (Salerno *et al.*, 2012; Salerno and Robbins, 2013). Simulations of the tensile deformation of silica nanofibers also show power-law-distributed strain bursts with scaling exponents differing from mean-field theory (Bonfanti *et al.*, 2018).

Experiment and simulations show that plasticity in amorphous solids typically initiates by localized plastic instabilities associated with the rearrangement of a small set of particles (also known as shear transformations (Argon and Kuo, 1979; Falk and Langer, 1998)) which releases the stored elastic energy (see Fig. 6.11(a)). These particle reorganization events have been observed and characterized in silica glasses (Horbach *et al.*, 1996; Coslovich and Pastore, 2009; Huang *et al.*, 2013; Bonfanti *et al.*, 2019), metallic glasses (Argon, 1979), colloidal glasses (Chikkadi *et al.*, 2011), foams (Dennin,

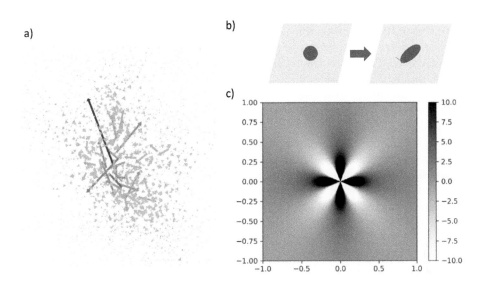

Fig. 6.11 (a) The deformation pattern associated with a localized plastic event in a simulation of deformed silica glass (Bonfanti *et al.*, 2019). (b) In the Eshelby inclusion problem a circular element of the elastic material is deformed into an ellipse and then placed back into the material (Eshelby, 1957). (c) The strain field produced by an Eshelby inclusion displays a characteristic quadrupolar structure.

2004), bubble rafts (Argon, 1979), and emulsions (Hébraud *et al.*, 1997; Clara-Rahola *et al.*, 2015). The rearrangement of the particles induces long-range stress fields in the material which can then cause further plastic instabilities, creating the observed widely distributed strain bursts. Notice the similarity between this process and the strain bursts caused by the motion of individual dislocations.

The stress fields associated with a shear transformation can be estimated assuming that the rearrangement is equivalent to the removal of a small ellipsoid inside the material which is then sheared and reinserted inside the elastic medium (see Fig. 6.11(b)). The problem is then mapped to the stress field produced by an inclusion in an infinite elastic medium, solved by Eshelby (1957). In the simple case of a two-dimensional infinite medium, the elastic stress field can be computed using the Green's function that in this case decays as

$$G(r, \theta) \propto \frac{\cos 4\theta}{r^2}, \tag{6.35}$$

as illustrated in Fig. 6.11c, showing the characteristic quadrupolar symmetry of the Green's function.

Since plastic deformation can be decomposed into a sequence of discrete localized plastic events, it is possible to devise mesoscale models based on this general mechanism (Baret *et al.*, 2002; Talamali *et al.*, 2011; Budrikis and Zapperi, 2013; Budrikis *et al.*, 2017; Nicolas *et al.*, 2018). In two-dimensional shear, the deformation is described by a single stress component and the problem becomes scalar. Following Talamali *et al.* (2011, Budrikis and Zapperi (2013), we consider a two-dimensional lattice of size $L \times L$ and associate to each site a plastic strain variable γ_i and a plastic stress threshold t_i. The system is initialized with $\gamma_i = 0$ at all sites and t_i is taken from a uniform distribution over [0,1). The external shear stress f is slowly increased until the stress on the weakest site reaches its threshold. At this point, the strain of the site is updated by increasing its value by a fixed amount $d\gamma$ while a new random yield threshold is selected from the same distribution. After the site has been strained, the stress on the other sites is recomputed using the Green's function. This may lead to other sites becoming unstable, leading to an avalanche. The size of the avalanche is the total strain increase that occurs before the external stress is increased again, and the duration is the number of time steps taken.

This simple two-dimensional model can be generalized to a fully tensorial model in two and three dimensions by generalizing classical rate-independent continuum plasticity with a $J2$/Von Mises type yield criterion to account for structural randomness and localized plastic events (Budrikis *et al.*, 2017). In practice one considers a lattice composed of D dimensional cubic elements with a local stress tensor σ. The yield condition is defined as

$$\Sigma_{eq} = \sqrt{(3/2)\boldsymbol{\Sigma}'\boldsymbol{\Sigma}'} > \Sigma_y, \tag{6.36}$$

where $\boldsymbol{\Sigma}' = \boldsymbol{\Sigma} - (1/D)\mathrm{Tr}\boldsymbol{\Sigma}\boldsymbol{I}$ is the deviatoric part of the stress tensor and $\boldsymbol{I}$ is the rank-2-unit tensor in D dimensions. In analogy with the scalar model, plastic events occur when the local equivalent stress exceeds a randomly assigned threshold Σ_y drawn from a uniform distribution over $[0, 1)$. Once an element fails, the local plastic strain ϵ^P is increased by an amount $\Delta\epsilon^P = \Delta\epsilon\hat{\epsilon}$, where the strain direction $\hat{\epsilon} = \boldsymbol{\Sigma}'/\sigma_{eq} = (2/3)\partial\sigma_{eq}/\partial\boldsymbol{\Sigma}'$ is chosen to maximize the locally dissipated energy while $\Delta\epsilon$ is a fixed strain increment. As for the scalar model, new random thresholds σ_y drawn from the same distribution are assigned to failed elements.

To solve the elastic–plastic problem, the stress field $\boldsymbol{\Sigma}$ has been obtained from the plastic strain field ϵ^P and the external boundary conditions. Two different methods can be used depending on the boundary conditions. For periodic boundary conditions, one can use the Green's function method as for the scalar model. In Fourier space, we can write

$$\Sigma_{ij}^{int}(\boldsymbol{q}) = G_{ijkl}(\boldsymbol{q})\epsilon_{kl}^{P}(\boldsymbol{q}). \tag{6.37}$$

where the Green's tensor $\boldsymbol{G}$ is obtained by treating the plastic strain as a sum of Eshelby inclusions of vanishing volume located at the center of each element. As discussed above, the stress field is known analytically for an infinite system (Eshelby, 1957) and the solution can easily be extended to a periodic system of size L. A spatially homogeneous "external" stress field $\boldsymbol{\Sigma}^{ext}$ arising from remote boundary tractions applied to the infinite contour can then be superimposed to the solution. In the case of open boundary conditions with assigned loading, the Green's tensor cannot be obtained analytically and one should use instead finite element calculation to compute it (Budrikis *et al.*, 2017).

Numerical simulations of the model in two and three dimensions under different deformation modes (uniaxial tension with free surfaces and applied tensile tractions, pure

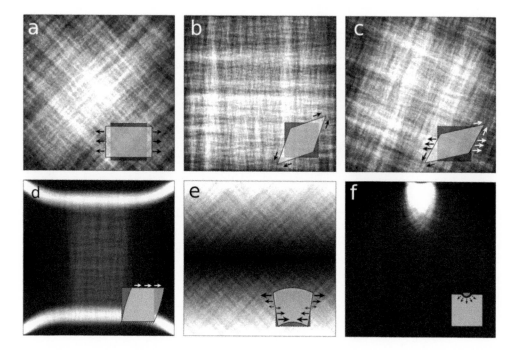

Fig. 6.12 Strain localization patterns obtained from the simulation of the two-dimensional tensorial model of amorphous plasticity. The loading conditions are (a) pure tension; (b) pure shear; (c) biaxial loading; (d) simple shear; (e) bending; (f) indentation. Source: Budrikis *et al.* (2017), CC-4.0 license.

shear, bi-axial deformation with applied tensile and shear traction and indentation) give rise to strain localization consistent with previous experimental and molecular dynamics studies (Vaidyanathan *et al.*, 2001; Ramamurty *et al.*, 2005; Shi and Falk, 2007; Maloney and Robbins, 2009; Chen and Lin, 2010). As illustrated in Fig. 6.12, the plastic strain field organizes into shear bands following the directions of maximum shear stress. For uniaxial tension, this is at approximately 45° to the tensile axis. In simple shear, the vertically fixed surfaces cause strong stress concentrations and strain localization in the corners of the sample. Under simulated 2D indentation with a circular indenter, strain localizes into a pattern of intersecting circles. This pattern is typical of shear bands observed in indentation of bulk metallic glasses (Su and Anand, 2006), and the simulations correctly reproduce them (Budrikis *et al.*, 2017). Clearly these types of strain patterning cannot be obtained using simpler scalar plasticity models.

Simulation results also revealed a strong universality of the avalanche exponent $\tau = 1.28 \pm 0.005$ which is not only independent of dimensionality (2D vs. 3D) but also does not depend on the description of stress (scalar vs. tensorial) nor whether the stress field is homogeneous and uniaxial (pure shear, pure tension) or inhomogeneous and biaxial near the boundaries (simple shear). Finally, results for pure tension and pure shear are the same, which indicates that the avalanche exponent is influenced

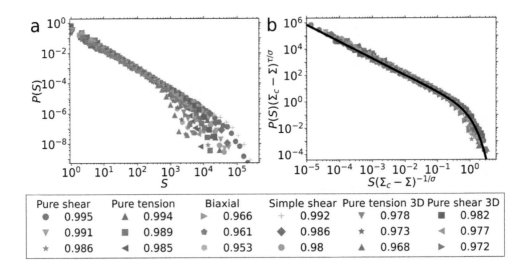

Pure shear		Pure tension		Biaxial		Simple shear		Pure tension 3D		Pure shear 3D	
●	0.995	▲	0.994	▶	0.966	+	0.992	▼	0.978	■	0.982
▼	0.991	■	0.989	✦	0.961	◆	0.986	★	0.973	◀	0.977
★	0.986	◀	0.985	●	0.953	●	0.98	▲	0.968	▶	0.972

Fig. 6.13 (a) Avalanche size obtained from the simulation of the tensorial model of amorphous plasticity in two and three dimensions with different loading conditions and different stress levels. (b) Data at different strain levels have been rescaled so as to obtain a collapse of the data. Source: Budrikis *et al.* 2017, CC-4.0 license.

neither by boundary conditions nor by the orientation of the simulation grid (see Fig. 6.13(a)). The size $S_{\max}$ of the largest avalanches diverges like $S_{\max} \propto (\Sigma_c - \Sigma)^{-1/\sigma}$ as the system average of the equivalent stress Σ approaches a non-universal critical value Σ_c. Accordingly, the distributions exhibit the scaling form $P(s) = u^{-\tau}\Psi(u)$ where $u = s/S_{\max}$ and the scaling function

$$\Psi(u) = \frac{A}{2\sqrt{\pi}} \exp\left(C\sqrt{u} - \frac{B}{4}u^\delta\right) \tag{6.38}$$

has been derived in the context of interface depinning (see also Eq. 4.53 and Le Doussal and Wiese (2012)). Using this scaling function all the distributions can be collapsed into a unique curve as shown in Fig. 6.13b. The exponent $\sigma = 1.95 \pm 0.01$ exhibits the same strong universality as the avalanche exponent τ. A third universal exponent γ_{st} defines the relationship between avalanche duration T and avalanche size, $s \propto T^{\gamma_{st}}$. To determine this exponent we define the duration of an avalanche as the number of simultaneous updates required from the start of the avalanche to the moment when all elements are stable again. Also in this case, simulations yield strong universality with an exponent $\gamma_{st} = 1.8 \pm 0.01$ for all loading conditions, including bending and indentation (Budrikis *et al.*, 2017).

6.7 Marginal stability

To understand the fundamental origin of plastic strain bursts, a central point to clarify is the way excitations are triggered by the external stress. In the context of amorphous plasticity, an important quantity in this respect is provided by the distribution $P(x)$ of the stresses x needed to induce a local excitation in the system. It has been proposed that this excitation distribution should scale as $P(x) \propto x^\theta$ in the $x \to 0$ limit, where the exponent $\theta > 0$ characterizes the system (Karmakar *et al.*, 2010). This scaling law was verified in molecular dynamics simulation of amorphous solids (Karmakar *et al.*, 2010; Hentschel *et al.*, 2015), mesoscale scalar elasto-plastic models (Lin *et al.* 2014a, 2014b, Liu *et al.* 2016; Budrikis *et al.* 2017), and in mean-field spin glasses (Pázmándi *et al.*, 1999; Yan *et al.*, 2015). It represents a signature of marginal stability, since the system can become locally unstable upon arbitrarily small external perturbations, and should encode crucial information about the properties of the ensuing crackling noise (Müller and Wyart, 2015). Notice that in classical depinning problems of elastic interfaces in random media $P(x)$ is flat for small x and $\theta = 0$. Hence, the marginally stable scaling with a characteristic "pseudogap" excitation spectrum is considered to be a distinguishing feature of glassy systems (Müller and Wyart, 2015).

It has been argued that the singular $P(x)$ is a consequence of the non-positive definite nature of the long-range interaction kernel mediating the collective deformation dynamics (Lin *et al.* 2014a, 2014b). This feature has been linked to the emergence of "extended criticality" in the crackling dynamics, when critical fluctuations take place over an extended range of control parameter values rather than only in the proximity of a "critical point" (Müller and Wyart, 2015) as reported earlier in simulations of a magnetic-field-driven mean-field spin glass (Pázmándi *et al.*, 1999). In the case of plastic deformation of a system with N degrees of freedom, one can relate the excitation spectrum to the size scaling of avalanche events: If the excitation spectrum scales as $P(x) \propto x^\theta$, the stress increment separating avalanches should scale as $N^{-1/(1+\theta)}$, and hence the number of events N_a within a stress interval $\Delta\sigma \sim 1$ scales as $N_a \sim N^{1/(1+\theta)} \ll N$. These few events must be responsible for an extensive strain increment, such that the mean avalanche size scales as $N/N_a \sim N^{\theta/(1+\theta)}$ and thus it diverges in the $N \to \infty$ limit.

It is possible to test the universality of the pseudogap scaling using the tensorial model of elasto-plasticity discussed above (Budrikis *et al.*, 2017). There, the local stability index can be defined as the normalized difference between the local applied stress and the local activation threshold, $x = \Sigma/\Sigma_t - 1$. As shown in Figure 6.14, $P(x)$ at criticality is described by the nontrivial exponent $\theta = 0.354 \pm 0.004$ for all loading conditions and dimensionalities. This observation corroborates the conjecture that a nontrivial yet universal local stability distribution is a generic signature of amorphous plasticity. This scaling is, however, only valid at the yield point and does not hold at lower stress. As the load increases, it initially follows the predictions of standard depinning theory (i.e. $\theta = 0$) and only the exponent increases to the asymptotic value $\theta = 0.35$ (Budrikis *et al.*, 2017). The crossover to the universal value is controlled by an effective coupling constant $C = E\Delta\epsilon/\langle\Sigma_t\rangle$, where $\langle\Sigma_t\rangle$ is the average yield threshold, $\Delta\epsilon$ is the strain increment, and E is the Young modulus. The behaviour in 2D and 3D simulations is similar: Away from the critical stress, the θ exponent takes the

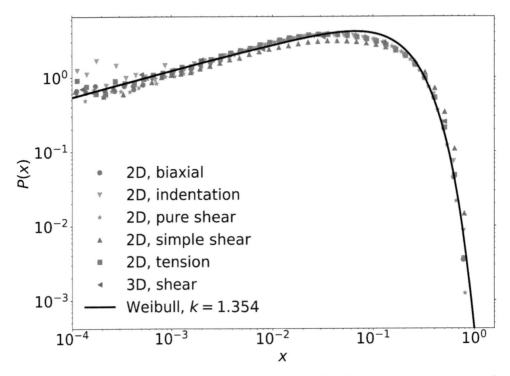

Fig. 6.14 Close to the critical stress Σ_c, the local stability distributions take a universal form with exponent $\theta = 0.35$ which is independent of the loading conditions. The full line shows a Weibull distribution with exponent $k = 1.35$ ($\theta = 0.35$), which is obtained from a simultaneous fit of all data sets. Source: Budrikis *et al.* 2017, CC-4.0 license.

trivial value $\theta = 0$, while very close to the critical stress, it crosses over to the non-trivial value $\theta \approx 0.35$ which is characteristic of the yielding state. Importantly, this cross-over occurs later for small values of the coupling constant $\mathcal{C}$, and in the limit of infinitesimally weak coupling $\mathcal{C} \to 0$, the system appears to flow toward depinning-like behavior ($\theta \approx 0$) except in close vicinity of the macroscopic yield stress.

To close this chapter, we notice that since dislocation systems have been shown to exhibit glass-like dynamics, originating from frustrated dislocation interactions (Bakó *et al.*, 2007), and extended criticality (Ispánovity *et al.*, 2014; Janićević *et al.*, 2015; Lehtinen *et al.*, 2016), it is expected that they also display non-trivial excitation spectra. Indeed this was confirmed by dislocation dynamics simulations in two and three dimensions (Ovaska *et al.*, 2017). Contrary to the case of amorphous materials, however, for dislocation systems the exponent θ is found to depend on dimensionality. The exponent also depends on the stress rate of the local perturbation, the presence or absence of quenched pinning, and whether the other dislocations not directly subject to the perturbation are allowed to move or not (Ovaska *et al.*, 2017).

7
Granular Matter

While granular media were originally thought to be the natural realization of self-organized criticality, it was soon realized that real sandpiles do not behave in the same way as sandpile models suggest. Instead of a power-law distribution of avalanches, one observes a prevalence of system-spanning events. This is due to the inertia of the grains, which can be reduced by using elongated grains (such as rice). The avalanches in the ricepiles closely follow the behavior observed in sandpile models. Avalanches are also found in granular media under shear, where, due to the complex grain arrangement and load transfer, one observes an intricate stick-slip behavior.

7.1   Granular matter: Statics and dynamics

Granular matter, an ensemble of discrete macroscopic particles, is found in several contexts, such as sand, seeds, and powders. For this reason, understanding its mechanics and dynamics has important applications in many fields, from industry to agriculture. We talk about granular media when the radius of the individual particles is larger than a few microns, so that thermal agitation is negligible. Thus, granular matter is inherently athermal and only the mechanical response plays a role. This is one of the main reasons why a theoretical treatment of granular media is a formidable problem: we have an ensemble of many particles, but we cannot directly apply the traditional tools of statistical mechanics and thermodynamics. Analogous to ordinary matter, granular media can be found in different "states" from the solid-like structure of cohesive granular assemblies, to the liquid-like response of a flowing granular bed to the gas-like state of a strongly vibrated powder. These granular states, however, are not easy to define unambiguously, due to the lack of a proper thermodynamic description. In addition, in many instances the granular assembly responds to external perturbations in a complicated way, combining together all these states. Think of a dune in which sand is transported by the wind in a gas-like form, flows over the slope as a liquid and remains solid-like in the bulk. Despite not having been investigated for several years, due to its importance and ubiquity granular matter has attracted a renewed interest in the physics community. Here, we would like to summarize some basic properties of granular matter which will be of interest for what follows. For a more detailed treatment, we refer the reader to various excellent review articles that have appeared in the literature (Jaeger *et al.*, 1996; Marone, 1998; de Gennes, 1999; Aranson and Tsimring, 2006; Herrmann *et al.*, 2013).

A static granular pack displays a number of remarkable features which are due to the way forces are transmitted by the grains. As first noticed by Janssen (1895), the pressure at the bottom of a granular column placed in a container is not a linear

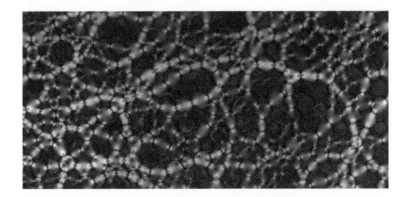

Fig. 7.1 Force chains in a granular material composed of photoelastic disks. Source: Daniels (2017) CC license 4.0.

function of the height, as would be the case for a regular solid or liquid. For a tall granular column, the pressure value is instead a constant, independent of height! The origin of this behavior is due to the contact forces between the grains, which create a macroscopic force component that is perpendicular to the side of the container. Due to the frictional forces acting on the container walls, a fraction of the normal load is transferred to the tangential direction, leading to a reduction of the normal pressure. Using simple assumptions for the normal and tangential stress components and basic friction laws, Janssen showed that the pressure profile has an exponential dependence on the height, saturating to a constant value (Janssen, 1895; de Gennes, 1999). Notice that this behavior has important implications, since it is responsible for catastrophic collapses occurring in silos when the wall pressure overcomes a rupture threshold.

Due to the disordered arrangement of the particles, contact forces in granular media follow intricate paths, giving rise to a network of force chains (Travers *et al.*, 1987; Liu *et al.*, 1995; Daniels *et al.*, 2017; Zadeh *et al.*, 2019). The structure of these networks can be revealed by using photoelastic particles (see Fig. 7.1), for which the refracting index depends on the state of deformation (Giusti *et al.*, 2016; Papadopoulos *et al.*, 2016). It is thus possible to directly highlight the stressed particles in the assembly, leading to spectacular images, such as the ones taken by the group of Behringer in a two-dimensional rotating cell where one can see the formation and breakup of force chains as the cell is sheared (Miller *et al.*, 1996; Howell *et al.*, 1999; Veje *et al.*, 1999). Transmission of forces through this irregular network is responsible for many peculiar features of granular mechanics, such as the value of the normal pressure being independent of height (as discussed above), but also of stress fluctuations and the tendency of the medium to *dilate* under shear (Reynolds, 1885; Bagnold, 1966).

The mechanical response of granular matter has been investigated for many years, due to the relevance of this problem for engineering and geophysics. The basic phenomenology is often described in terms of *soil mechanics*, which has become a discipline in itself (Taylor, 1945; Wood, 1990). An idealization of a shear test is described in Fig. 7.2, illustrating a granular medium sheared by a deviatoric stress σ under a normal pressure p. The yield stress for granular matter can be expressed in analogy

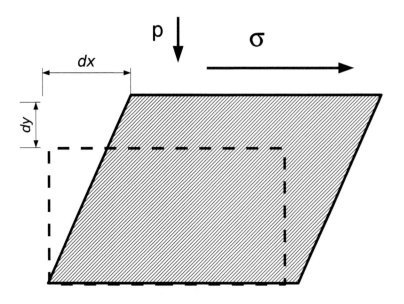

Fig. 7.2 An idealized view of the shear of a granular medium.

with frictional sliding as

$$\sigma_c = \mu p + C, \tag{7.1}$$

where μ is the friction coefficient and C is a cohesion parameter. The response of the medium also depends on the density of the granular assembly: A dense medium will dilate under shear, while a loose one will compact. Experiments show that a critical density ρ_c exists at which the granular medium flows steadily, leading to the notion of a *critical state*. Clearly the critical state of soil mechanics is not necessarily related to critical phenomena in phase transitions, but in some cases shearing a granular medium produces scaling laws, as we will discuss in the following. Here, we would like to discuss a simple argument proposed by Taylor (1945) to describe dilatation of a granular medium. The idea is that in order to be sheared, the medium should also dilate to overcome the constraints of interlocked grains. As shown in Fig. 7.2, under a shear stress σ and pressure p the medium deforms by dx and dy in the horizontal and vertical directions, respectively. One can then equate the mechanical work to the frictional work obtaining

$$\sigma A dx - p A dy = \mu p dx, \tag{7.2}$$

where A is the sample area. From Eq. 7.2, we obtain a relation between the shear stress σ_c, the dilatancy angle ψ, defined by $\tan \psi \equiv dy/dx$, and the friction angle θ, defined by $\tan \theta \equiv \mu$:

$$\sigma_c = p(\tan \theta + \tan \psi). \tag{7.3}$$

This equation can also be expressed as a relation between friction coefficients and the dilatancy as (Marone, 1998)

$$\mu_c = \mu + \frac{dy}{dx}, \tag{7.4}$$

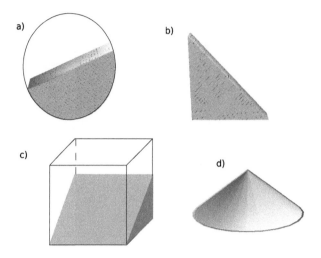

Fig. 7.3 The most common setups used to study granular avalanches: (a) the rotating drum, (b) the one-dimensional pile, (c) the rectangular pile, and (d) the conical pile.

where μ_c is the measured friction coefficient of the granular medium. The soil mechanics description of granular shear is usually more complicated than the one we have sketched here. Granular friction is a very complex phenomenon that is difficult to capture entirely with simple phenomenological laws and a vast literature has been devoted to this topic. We will return to it in section 7.4

7.2   Avalanches in granular piles

The original paper by Bak *et al.* (1987) introducing the sandpile model of SOC stimulated several experimental groups to attempt an experimental realization of this phenomenon by studying the avalanches in a pile of sand (Jaeger *et al.* 1989; Evesque and Rajchenbach 1989; Held *et al.* 1990; Bretz *et al.* 1992; Rosendahl *et al.* 1993; 1994). The task was ultimately very difficult to accomplish and several conflicting results appeared in the literature over the years. Avalanches were studied in several experimental setups, using different types of granular materials, such as glass or metal beads and even rice. We summarize in Fig. 7.3 the most common setups employed in experiments: the rotating drum, the one dimensional pile, the rectangular pile, and the conical pile. Jaeger *et al.* (1989) measured the sand flowing down using a capacitor at the base of a rectangular pile. They found that avalanches displayed Gaussian statistics with well-defined characteristic sizes. Essentially the dynamics was dominated by large system-spanning avalanches occurring periodically. Similar oscillations were also reported for a rotating drum (Jaeger *et al.*, 1989; Evesque and Rajchenbach, 1989).

Other experimental investigations of avalanches in granular piles were performed using a conical pile, placed on a mass balance to measure avalanche fluctuations (Held *et al.* 1990; Rosendahl *et al.* 1993; 1994). The results by Held *et al.* (1990) suggested

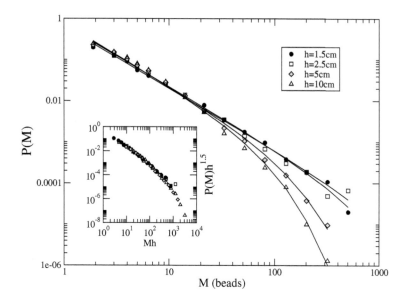

Fig. 7.4 The distribution of avalanche sizes for the bead pile experiment of Costello *et al.*. The distributions are collected for different values of the dimension drop height h and can then be collapsed (inset). Source: Data from Costello *et al.* (2003).

that only small piles would display finite-size scaling in the avalanche distribution. In particular the avalanche size distribution can be scaled as

$$P(M, R) = g(M/L^\nu)/L^\beta, \tag{7.5}$$

where M is the mass of the avalanche, L is the base diameter, $\nu \simeq 0.9$ and $\beta \simeq 1.8$. The scaling function $g(x)$ can be approximated by a stretched exponential. For large piles, the scaling breaks down and periodic large avalanches were observed instead. Similar results were also found by Rosendahl *et al.* (1993, 1994).

A more systematic study of avalanches in conical piles was performed by Costello *et al.* (2003). Much larger beads $d = 3$mm were employed in this experiment, thus avoiding cohesive effects due to humidity. In addition, those conducting the experiments glued a set of beads on the base of the pile, testing the effect of different arrangements, and varied the base diameter from $L = 30d$ to $L = 90d$. Finally, the effect of the drop height was systematically studied. The result was a power-law distribution of avalanche sizes, measured as the mass variation M in the pile, over two decades, with an exponent close to $\tau_M = 1.5$. The cutoff of the distribution was found to be independent of the bead arrangement on the base and of its shape, and slightly dependent on the base diameter. It was, however, strongly dependent on the drop height h, allowing for a data collapse according to

$$P(M, h) = M^{-\tau_M} f(Mh), \tag{7.6}$$

with $\tau_M = 1.5$. Thus, the cutoff increases as h is decreased (see Fig. 7.4).

Large spanning events appear to be an unavoidable consequence of grain inertia, combined with a velocity-weakening friction law. It is possible, however, to considerably reduce this effect by forming a pile with rice. Frette *et al.* (1996) studied a two-dimensional ricepile and found a power-law distribution, over almost two decades, when using elongated rice grains. The avalanche exponent was estimated to be $\tau = 2$. Round grains rolled instead of sliding, yielding periodic large events, as commonly observed with beads. Notice that the size s of an avalanche in this experiment was defined as the integral of the difference between the profiles of the pile after the addition of one grain. This definition of size is different from the one employed in many other experiments, which is usually the mass flowing over the rim. The problem of different definitions is sometimes overlooked when one compares the experiments to the theory.

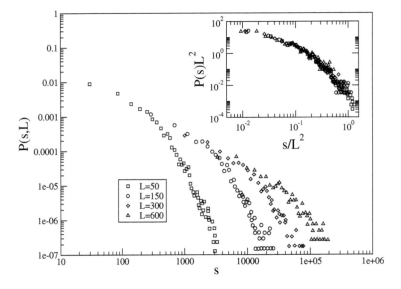

Fig. 7.5 The distribution of avalanche sizes for the ricepile experiment of Aegerter *et al.*. The distributions are collected over different windows of dimension L and can then be collapsed (inset). Source: Data from Aegerter *et al.* (2003).

Avalanches in ricepiles were systematically studied in three dimensions by the group of Wijngaarden (Aegerter *et al.*, 2003). Using an optical setup, it was possible to reconstruct the entire surface of the pile with a resolution of less than a single grain. This allowed the precise monitoring of the roughening of the granular surface and the associated avalanches. Comparing different images of the piles, they could estimate the displaced size s in the pile and compute histograms. The distributions were obtained for different image resolutions, varying the window size L (see Fig. 7.5). According to Aegerter *et al.* (2003), the data could be collapsed using the scaling form

$$P(s, L) = s^{-\tau} f(s/L^D), \tag{7.7}$$

with $\tau = 1.2$ and $D = 2$. The collapse with this form is poor, but a better one could be obtained by a simpler form (see the inset of Fig. 7.5) already proposed by Held *et al.* for conical piles (see Eq. 7.5 and Held *et al.* 1990):

$$P(s, L) = L^2 g(s/L^2). \tag{7.8}$$

Notice that the data do not show a clear power law, which could be seen only in the envelopes of the curves, while the scaling function $g(x)$ resembles, again, a stretched exponential.

In summary, the avalanche behavior in real granular material appears to be quite different from the idealized behavior of SOC sandpile models. The reason for this behavior can be traced back to the frictional instabilities occuring in granular materials (Nagel, 1992). A granular pile is characterized by two important angles: the angle of repose θ_c and the angle of maximal stability θ_m. As we increase its slope, the pile remains stable until we reach an angle θ_m where an instable flow occurs, reducing the angle to θ_c. As first pointed out by Bagnold (1966), the difference $\delta = \theta_m - \theta_c$ is due to dilatancy: The superficial layers expand before flowing, leading to a reduction of the friction force and hence to an instability. The behavior of the pile resembles more that of a first-order transition rather than a second-order critical point.

Jaeger *et al.* (1990) have incorporated these observations in a simple friction law that can qualitatively reproduce the results. The motion of a grain of sand along the pile follows Newton's law

$$ma = mg \sin\theta - F_f, \tag{7.9}$$

where F is a velocity-dependent friction force. From some simple considerations on the energy lost by the grain in motion, the friction force can be written as a function of the velocity v as

$$F_f(v) = \frac{Amg}{1 + Bv^2/gd} + Cmg(v^2/gd), \tag{7.10}$$

where d is the grain diameter and A, B, and C are numerical constants. The friction force in Eq. 7.10 decreases for velocities smaller than $v_0 \simeq (gd)^{1/2}$ and then increases again. The velocity weakening is responsible for the unstable stick–slip behavior, giving rise to periodic large avalanches. We can then obtain the angle of maximal stability from the static friction $\sin\theta_m = F(0)/mg$ and the angle of repose from $\sin\theta_c \simeq F(v_0)/mg$.

The avalanche scaling observed for small piles (Held *et al.*, 1990) was attributed by Nagel to the small value of $\theta_m - \theta_r \simeq 2°$ (Nagel, 1992). If the base of the pile is too small, adding a single grain would change the slope from below θ_r to above θ_m. Hence the experimental conditions would not be able to distinguish between the two angles, preventing the large-scale oscillatory behavior. An estimate for the base diameter at which the scaling will break up yields $L > 30d$ (Nagel, 1992) which is in agreement with the experiments of Held *et al.* (1990). This crossover behavior, however, was not observed in the experiments of Costello *et al.* (2003), possibly because of a very small value of $\theta_m - \theta_r$ in those experimental conditions. Finally, scaling is more convincing in ricepiles where one could reasonably assume that inertial effects are negligible. Nevertheless, the experimental results did not find a convincing quantitative theoretical explanation. Indeed the numerical results obtained for one-dimensional 'ricepile models" are not in agreement with the experiments.

7.3 Stick–slip motion in granular shear

As we discussed in section 7.1 the shear deformation of granular matter has been the object of intense study in the soil mechanics, geophysics, and physics communities. Here, we would like to discuss some interesting properties observed in the stress fluctuations of deformed granular media. These arise quite naturally as a result of the intricate force chains we have discussed previously. As granular matter deforms, grains displace, which leads to avalanche-like rearrangements of the force chains that result in macroscopic stress fluctuations. Several experiments and models have been devised in the past to understand this phenomenon.

The frictional sliding of a granular bed has been studied in a series of experiments by the group of Gollub (Nasuno *et al.* (1997, 1998); Géminard *et al.* 1999; Losert *et al.* 2000). The apparatus consisted of a transparent plate pulled by a spring over a granular bed, which in some experiments was also placed under water (Géminard *et al.*, 1999). The plate surface was roughened by gluing a layer of particles onto it. The experiments allowed one to follow the horizontal x and vertical y positions of the plate as well as the deflection of the spring δx. Knowing the mass of the plate M and the sring stiffness k it was then possible to extract the effective friction force from

$$M\ddot{x} = k\delta x - F_f, \tag{7.11}$$

where $\delta x = Vt - x$ depends on the velocity V at which the spring is pulled. For slow driving velocities V, the motion exhibited a periodic stick–slip, crossing over to continuum sliding at higher V. The stick–slip motion became more irregular when the velocity was lowered even more, but this regime was not studied in detail. By recording the friction force during a single slip event, it was possible to show that the force was not simply a function of the instantaneous velocity, as for instance in Eq. 7.10 (Jaeger *et al.*, 1990). In particular, it was observed that the frictional force performed a hysteresis loop as a function of the instantaneous velocity (see Fig. 7.6). This observation reflected the dependence of the friction force on other internal (or state) variables. In particular, it was shown that a slip event is associated with a vertical displacement of the slider, indicating that dilatancy is playing a role, as expected from general considerations. For wet granular media, the effective friction coefficient could not be described by the simple relation in Eq. 7.4 since it was found to depend on the dilatancy rate (Géminard *et al.*, 1999).

The same experiments revealed some additional dynamical effects: In the stick phase, before a slip event, the plate was observed to slowly creep. The creep motion was found to accelerate in "anticipation" of the slip event, acting as a sort of precursor. Considering the fact that thermal activation is negligible for granular media, the creep regime was probably due to vibrations in the apparatus or other mechanical perturbations. The stick phase is also characterized by an aging of contacts: The friction coffiecient increases with the logarithm of the stick time (Losert *et al.*, 2000). This behavior is common in solid-on-solid friction, but here it was observed only if a shear stress was applied during the waiting time.

The role of internal rearrangements in the shear deformation of granular media was better clarified by two-dimensional experiments (Veje *et al.*, 1999; Howell *et al.*, 1999; Geng and Behringer, 2005). A two-dimensional Couette cell was filled with photoelastic

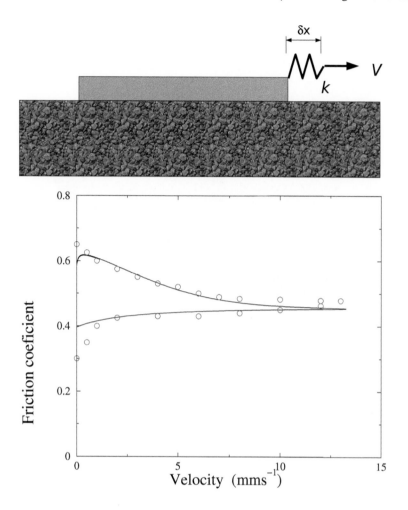

Fig. 7.6 (Top) The experimental apparatus of the granular friction experiments performed in Nasuno *et al.* (1997, 1998); Géminard *et al.* 1999; Losert *et al.* 2000. (Bottom) The frictional coefficient $\mu = F_f/Mg$ of the granular bed as a function of the instantaneous velocity during a slip event. The line is the prediction of the rate-and-state model studied in Lacombe *et al.* (2000).

disks and shear was applied rotating the inner cylinder at constant angular velocity. In this way, it was possible to show that the shear deformation is concentrated on a small shear band of few particles, with a velocity profile decaying roughly exponentially from the inner radius (Veje *et al.*, 1999). The photoelastic property of the disks enabled the witnessing of the formation and destruction of force chains as the medium was sheared (Howell *et al.*, 1999). The intensity of the light transmitted by the disks was converted into a force, allowing researchers to study its distribution (Daniels and Hayman, 2008). The properties of the medium were shown to depend very strongly on the particle

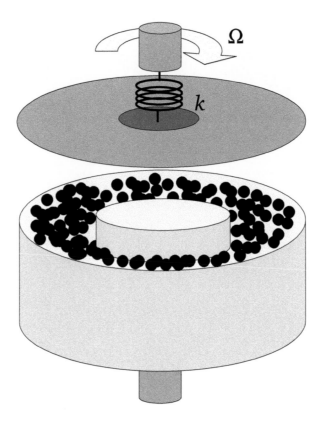

Fig. 7.7 The experimental apparatus of the granular friction experiments performed in (Dalton and Corcoran, 2001; Dalton and Corcoran, 2002; Baldassarri *et al.*, 2006).

density, or packing fraction. Varying this parameter, the system undergoes a transition, changing several properties: The compressibility increases, the mean velocity decreases, and the force chains and the stress distribution change their characters (Howell *et al.*, 1999).

A similar experimental setup, but in three dimensions, was used in Dalton and Corcoran (2001, 2002); Dalton *et al.* 2005; Baldassarri *et al.* 2006 – a Couette cell made of a circular channel containing mono-dispersed 2-mm glass beads (see Fig. 7.7). An annular plate was driven over the top surface of the channel by a motor through a torsion spring with stiffness k. To ensure a granular shearing plane, the annular top plate had a layer of beads glued to its lower surface and the system was initialized by

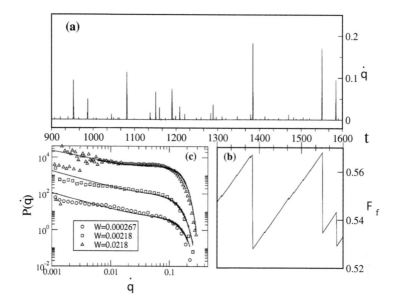

Fig. 7.8 The signal measured in the experiments of Baldassarri *et al.*. (a) instantaneous velocity of the disk; (b) instantaneous torque exerted by the spring. (c) Velocity distributions for different values of Ω. Source: Baldassarri *et al.* (2006).

turning it at a slow velocity Ω for a long time, in order to approach a stationary state. The device measured the angular position θ of the top plate and the deflection of the torsion spring. From this, with knowledge of the torsion spring constant k and the moment of inertia of the plate I, one could obtain angular velocity and acceleration, instantaneous torque, and work done by the spring. For slow values of the driving angular velocity Ω the system displays an irregular stick–slip regime. In analogy with Eq. 7.11, the equation of motion of the disk reads:

$$I\ddot{\theta} = k(Vt - \theta) - \Phi_f, \tag{7.12}$$

where Φ_f is the frictional torque exerted by the medium.

The typical instantaneous velocity signal displayed a very erratic behavior with pulses of widely fluctuating magnitudes (see Fig. 7.8), and its probability density distribution was found to be a power law with cutoff, spanning more than two decades. The presence of power-law statistics in the stick–slip phase of a sheared granular was first reported some years ago in slightly different experiments (Dalton and Corcoran 2001, 2002). In order to get a more complete characterization of the fluctuations and their temporal scales, one can analyze duration T and a size S associated with each pulse. In this case S is equal to the angular slip of the plate during the pulse. The distributions of T and S, measured for different values of the driving angular velocity Ω, are reported in Fig. 7.9 (Baldassarri *et al.*, 2006). They display a complex shape, in which an initial decay is followed by a characteristic peak at larger scales, indicating broad fluctuations over many scales. Note that the position of the peaks does not depend on the driving velocity Ω.

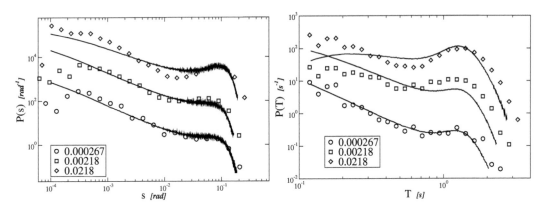

Fig. 7.9 Slip size (left) and duration (right) probability distributions, for different values of the driving angular velocity. The lines are the results for the numerical integration of Eqs. 7.12-7.18. The distributions for different drives are vertically shifted for clarity. Source: Baldassarri *et al.* (2006).

7.4 Modeling granular avalanches

The friction law derived in Jaeger *et al.* (1990) explains some basic features of granular avalanches, but it appears to be inadequate to completely describe the frictional properties of granular media when compared to other types of experiments. In particular, the Eq. 7.10 would predict a stick–slip behavior for $V < v_0$, but it could never reproduce the hysteresis shown in Fig. 7.6. To this end, one should consider the *rate-and-state* model in which the friction force is considered as a function of the velocity and of other variables, describing the state of the medium. In the context of solid-on-solid friction, one typically employs the 'age" of the frictional contacts as a state variable ϕ. This variable is equivalent to the time in the stick phase and depends on the velocity in the slip phase. In particular, ϕ evolves according to (Marone, 1998)

$$\frac{d\phi}{dt} = 1 - v\phi/D, \tag{7.13}$$

where v is the instantaneous velocity and D is a characteristic length scale. The frictional force is chosen to be of the type (Marone, 1998)

$$F_f(v, \phi) = F_0 - a \log v + b \log \phi, \tag{7.14}$$

where a and b are two parameters. According to Eq. 7.14, the static friction increases logarithmically with time in the stick phase and decreases with the velocity in the slip phase. Rate-and-state equations have also been adapted to granular media where the natural state variable is provided by a measure of dilatancy. For instance, in Lacombe *et al.* (2000) the frictional force is chosen to depend on the vertical position of the slider y:

$$F(v, y) = F_d - \beta \frac{y - y_m}{R} - \alpha v \frac{y - y_m}{R}. \tag{7.15}$$

Here, the first two terms give the static friction force ($v = 0$) as function of y. The velocity dependence is linear, mediated by the factor $y - y_m$ which vanishes when the bed is fully dilated (i.e. $y = y_m$). The vertical displacement of the granular medium during shear evolves as

$$\dot{y} = -\frac{y}{\eta} - v\frac{y - y_m}{R}. \tag{7.16}$$

By numerically solving this model, it is possible to reproduce the hysteresis property of the force velocity curve during slip (see Fig. 7.6) as well as other features of the experiment (Lacombe *et al.*, 2000).

The models described above do not contain any form of randomness and are therefore not adequate to reproduce the fluctuation in the slip velocities observed in experiments. To this end, it was proposed to proceed in the spirit of rate-and-state models, but introducing a fluctuating component of the friction force. This was done in Baldassarri *et al.* (2006) for the experimental setup described in Fig. 7.7. The friction torque was separated in two parts $\Phi_f = \bar{\Phi}_f(\dot{\theta}) + d\Phi_f(\theta)$, where the average velocity-dependent component is described by

$$\bar{\Phi}(v) = F_0 + \gamma(v - v_0 \ln(1 + v/v_0)), \tag{7.17}$$

where F_0 is the average static friction, v_0 corresponds to the minimum in the average torque, and γ is the high-velocity damping. Notice that Eq. 7.17 displays similar features to Eq. 7.10 and is found to accurately describe the experimental data.

The fluctuating part of the friction torque is a direct consequence of the disordered structure of the force chains present in the granular medium. As the disk slips by a small angle $\delta\theta$, the friction torque is increased or decreased by a random amount, reflecting a small rearrangement in the structure force chains. On the other hand, large slips will lead to a complete rearrangement of the grains and thus to a decorrelation of the random torque. These observations were cast in mathematical terms assuming that the fluctuating friction torque follows the confined Brownian process:

$$\frac{d\Phi_f}{d\theta} = \eta(\theta) - F_f/\xi, \tag{7.18}$$

where η is an uncorrelated Gaussian noise term with variance Δ. The parameter ξ represents the typical correlation length as can be seen from the power spectrum which, for the process above, is given by

$$S(k) = \left\langle \left| \int d\theta F_f(\theta) \exp(-i\theta k) \right|^2 \right\rangle = \frac{2D}{\xi^{-2} + k^2}. \tag{7.19}$$

Once the parameters ξ, Δ, F_0, γ, and v_0 have been extracted from the experimental data, the numerical integration of the equation of motion results in a stochastic velocity signal to be compared with the experiments. The model was found to reproduce the complex phenomenology of the granular dynamics, although it is based on few average parameters, whose values are directly measured on the system in study. In particular, the velocity and avalanche distributions can be reproduced with good accuracy (see Figs. 7.8 and 7.9). It would be interesting to generalize this model, incorporating for instance the effect of dilatancy, to study the crossover from irregular to periodic stick–slip motions.

Further theoretical insight could come from numerical simulations of assemblies of granular particles. Models of this kind have been used in the past to reproduce the experimental results for granular avalanches, friction, and shear. From the technical point of view these simulations are very complicated because of the variety of effects present in these systems, such as anelastic collisions, friction, and elastic deformation. None of these processes are easy to capture correctly in a model. Two-dimensional simulations so far have reproduced some essential features of granular shear (Thompson and Grest, 1991; Schöllmann, 1999), but it remains necessary to establish a link with more phenomenological and tractable theoretical models.

8
The Barkhausen Effect

In 1919 H. Barkhausen discovered that the magnetization of iron is associated with a crackling noise that could be revealed as an induction pulse in a coil. This observation provided an indirect indication of the existence of ferromagnetic domains and stimulated intense research activity spanning eight decades. The Barkhausen effect is probably the cleanest example where the mechanism of self-organized criticality can be applied to an experiment. The scaling exponents describing Barkhausen avalanche distributions in bulk materials can be quantitatively explained by studying the depinning of domain walls. Our understanding is still not complete in the case of thin magnetic films where the domain structure and the avalanche dynamics are often quite intricate.

8.1 Statistical properties of Barkhausen noise

The magnetization process of iron and other ferromagnetic materials displays hysteresis and the magnetic state of the sample depends on the previous history. It was first revealed in 1919 by Barkhausen that magnetization changes in response to a changing field are not smooth but proceed in bursts of activity. In the original experiment, a magnet was displaced close to a piece of iron and flux changes were recorded by a coil wound around the sample. By connecting the coil to an amplifier, Barkhausen could hear a characteristic crackling noise that was then named after him. The Barkhausen effect is considered the first indirect evidence of the motion of ferromagnetic domains and has been widely investigated over the past century. A reliable statistical analysis of noise properties has been obtained only in the past decades, leading to a quite detailed understanding of the phenomenon. For a complete discussion of this topic, including an historical perspective, we refer to the review article by Durin and Zapperi (2005).

The Barkhausen noise is still measured by inductive methods as in the original paper: A pickup coil is wound around the sample which is then placed into a solenoid providing a varying magnetic field . Flux changes appear in the amplifier as a voltage signal that can then be statistically analyzed. Most experiments reported in the literature record the noise only in a small region around the coercive field. This is because the signal is stationary only in this region, as shown in Fig. 8.1, while non-stationary signals have strong effects on the statistics as discussed in Durin and Zapperi (2006).

The statistical properties of the Barkhausen noise are usually characterized by a set of distribution functions, typically displaying scaling over some regime. A sequence of pulses is extracted from the noise signal $V(t)$, introducing a low threshold defining the beginning and the end of each avalanche. The distance between these points defines the avalanche duration T, while the size s is the area below the pulse. Since the voltage is

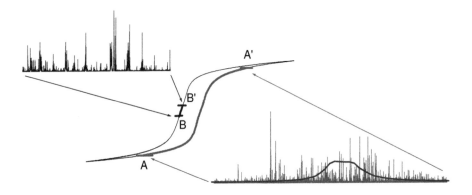

Fig. 8.1 Hysteresis loop of an $Fe_{85}B_{15}$ amorphous alloy under moderate tensile stress (10 MPa). The time derivative of the magnetization is Barkhausen noise. The signal detected between A and A' corresponding to a magnetization range of about 1.8 T, is not stationary, as revealed by the corresponding time signal in the bottom right corner. The signal measured around the coercive field, between B and B' is stationary, and corresponds to a range of about 0.1 T. Source: Durin and Zapperi (2006)

directly proportional to the flux change, the size s is proportional to the magnetization steps that we could see by magnifying the hysteresis loop curve.

The distribution of avalanche sizes and durations are usually described by

$$P(T) = T^{-\alpha}g(T/T_0) \tag{8.1}$$

and

$$P(s) = s^{-\tau}f(s/s_0), \tag{8.2}$$

where $g(x)$ and $f(x)$ are two cutoff functions decaying rapidly to zero around T_0 and s_0, and α and τ the critical exponents. As for other avalanche phenomena discussed throughout this book, the size is related to the duration by a scaling relation $s \sim T^\gamma$ which can be verified in practice by measuring the expectation value of the size of an avalanche of duration T.

As shown in Fig. 8.2, the recorded scaling exponents display a certain degree of *universality*: In amorphous alloys we have $\tau \simeq 1.3$ and $\alpha \simeq 1.5$ and in polycrystals $\tau = 1.5$ and $\alpha = 2$. The two 'classes" are also distinguished by the dependence on the field rate of the distributions (see Fig. 8.3): In amorphous alloys the exponents are rate independent but the cutoff function displays an increasing peak, while in polycrystals the exponent values decrease with the rate. This behavior is in line with the general arguments discussed in section 1.6 that predict a linear dependence only when $\alpha = 2$, which is indeed the value measured in polycrystalline samples.

The scaling of the cutoff of the distributions was investigated in detail (Durin and Zapperi, 2000) by a systematic change of the demagnetizing factor k related to the sample apparent permeability μ_{app} by the expression $\mu_{app} = \mu/(1 + k\mu)$, with μ the intrinsic permeability. Avalanche distributions were measured in polycrystalline and

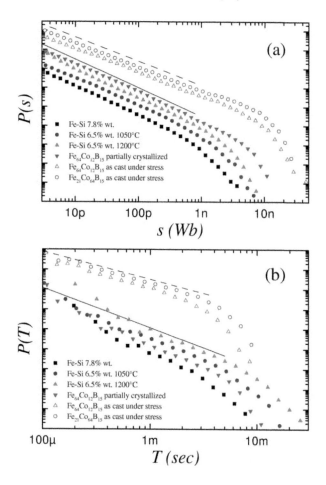

Fig. 8.2 (a) Distributions of Barkhausen jump sizes measured in different materials. The solid line has a slope $\tau \sim 1.5$ while for the dashed one $\tau \sim 1.27$. (b) Similar plot for duration distributions. The solid line has a slope $\alpha \sim 2$, while for the dashed one $\alpha \sim 1.5$. Source: Durin and Zapperi (2000), Fig. 1.

amorphous samples that were progressively cut to change the apparent permeability μ_{app}, and thus calculated the demagnetizing factor k. In this way it was possible to verify that the cutoffs s_0 and T_0 strongly depend on k, as in Fig. 8.4, and follow the scaling relations $s_0 \sim k^{-1/\sigma_k}$, and $T_0 \sim k^{-\Delta_k}$. Best-fit estimation of the exponents gives $1/\sigma_k \sim 0.57$, $\Delta_k \sim 0.30$ for the polycrystalline class, and $1/\sigma_k \sim 0.79$, $\Delta_k \sim 0.46$ for the amorphous one.

Table 8.1 summarizes the exponent values reported in the literature considering only those obtained using a reliable statistical sampling, leaving out, for instance, the results of the papers of Cote and Meisel (Cote and Meisel, 1991; Meisel and Cote, 1992), who were the first to suggest the relation between the Barkhausen effect and self-organized criticality (Bak *et al.*, 1987). Most, but not all, the exponents can be

Ref.	Material	Type	τ	α	$1/\sigma\nu z$	Rate dep.
(1)	81%NiFe	Wire	1.73	2.28	1.63	Yes
(2)	Vitrovax 6025X	Ribbon	1.77	2.22	1.51	?
(3)	Perminvar	Strip	1.33	-	-	No
(4)	Annealed steel	Strip	1.27	-	-	No
(5)	$Fe_{64}Co_{21}B_{15}$, $Fe_{21}Co_{64}B_{15}$ amorphous under stress	Ribbon	1.3	1.5	~1.77	No
(6)	$Fe_{21}Co_{64}B_{15}$ amorphous (no stress)	Ribbon	1.46	1.74	1.70	No
(7)	SiFe 1.8%	Strip	1.5	2	-	Yes
(5)	SiFe 6.5%, SiFe 7.8%, $Fe_{64}Co_{21}B_{15}$ polycrytal	Ribbon	1.5	2	~2	Yes
(8)	$Fe_{85}Co_{15}$ (entire loop)	Ribbon	1.7	2	1.77	No
(8)	$Fe_{85}Co_{15}$ (coercive field) (coercive field)	Ribbon	1.38	1.65	1.77	No

Table 8.1 Experimental critical exponents reported in the literature which can be considered sufficiently accurate and reliable. τ, and α are the critical exponents of the size and duration distributions, respectively. The exponent $1/\sigma\nu z$ relates the average size of an avalanche to its duration. 'Rate dep." indicates whether the exponents τ and α depend on the applied field rate. Source: The experimental data are taken for the following references: (1)Lieneweg and Grosse-Nobis (1972); (2)Spasojevic *et al.* (1996); (3)Urbach *et al.* (1995a);(4)McMichael *et al.* (1993); (5)Durin and Zapperi (2000);(6)Mehta *et al.* (2002);(7)Durin *et al.* (1995a);(8)Durin and Zapperi (2006)

associated into one of the two aforementioned classes. Several results seem to suggest the presence of a third class with $\tau \simeq 1.7$ reported in Lieneweg and Grosse-Nobis (1972, Spasojevic *et al.* (1996), but this idea is called into question by noticing that those exponents are obtained in non-stationary conditions. It was shown in Durin and Zapperi (2006) that one can measure two different sets of the exponents in the same sample by recording the signal through the entire loop or just around the coercive field (see Fig. 8.1). Sampling the distribution over a non-stationary signal leads to a higher exponent (see Fig. 8.5), as already discussed in the context of fracture in Chapter 5. It is important to remark that the results reported in Table 8.1 refer to soft magnetic materials, for which domain wall motion is the main magnetization mechanism. Less is known about the noise statistics in hard magnets, where the magnetization structure is more intricate. In addition, all the samples considered are essentially three dimensional. Thin films will be briefly discussed in section 8.5

Most of the earlier literature on the Barkhausen effect was devoted to the analysis of the noise power spectrum. This was mainly dictated by the practical applications of soft magnetic materials where the control of the spectral properties has a key importance. In general, the Barkhausen spectrum $F(\omega)$ displays some common features at high frequency $F(\omega)$, has a typical $1/f^a$ shape with $a = 1.7 \div 2$, and has an amplitude that scales linearly with the average magnetization rate $\dot{M}$. At a frequency roughly proportional to $\dot{M}^{1/2}$, the spectrum displays a marked peak and at lower frequencies,

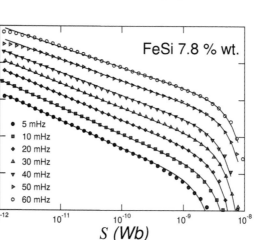

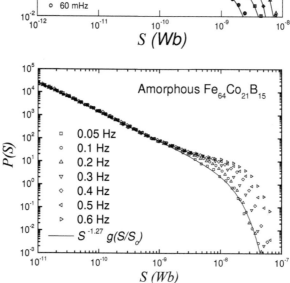

Fig. 8.3 Distributions of the avalanche sizes s for the polycrystalline SiFe (a) and the amorphous (b) material at different applied field frequencies. In the polycrystalline sample, the critical exponent τ varies linearly with the frequency (Durin *et al.*, 1995*a*; Durin *et al.*, 1995*b*) between 1.5 and 1.15, while it is independent in the amorphous one $\tau \sim 1.27$. Source: Durin and Zapperi 2005.

the spectrum scales as f^{ψ}, with $\psi \sim 0.6$, or ~ 1.

In more detail, single-crystal and polycrystalline SiFe strips display the simplest power spectrum, with $a \sim 2$, and $\psi \sim 0.6$. Polycrystalline SiFe ribbons at high Si content have the same exponents with some deviations. Amorphous materials under tensile stress show instead $a = 1.7 - 1.8$, and $\psi \sim 1$. Amorphous materials, where a proper annealing induces a partial crystallization, show a sort of intermediate case: The high-frequency part is more akin to the polycrystalline SiFe samples with $a \sim 2$, while $\psi \sim 1$ keeps the values of a pure amorphous material. This crossover is also reflected via a corresponding change in the avalanche exponents, which take the values measured

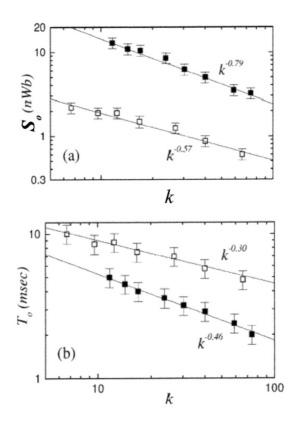

Fig. 8.4 The cutoff of the Barkhausen size (a) and duration (b) distributions as a function of the demagnetizing factor k in Fe-Si 6.5 wt% alloy (empty symbols) and an amorphous $Fe_{21}Co_{64}B_{15}$ sample under constant tensile stress (filled symbols). Source: Durin and Zapperi (2000).

in SiFe, as shown in Table 8.1. Despite these differences, all these materials have the same dependence of the peak frequency on $M^{1/2}$. As discussed in section 1.5, under some general conditions in avalanche phenomena the spectral exponent is related to the size–duration scaling exponent: $a = \gamma_{st}$. This prediction agrees very well with the experiments, as we can see in Fig. 8.6. The low-frequency parts of the spectra reflect the long time dynamics which is outside the scaling regime and at present is not understood by a simple scaling argument.

Finally it has been tested that the average avalanche shape scales in a universal way

$$v(t,T) = T^{\gamma_{st}-1}\mathcal{V}(t/T), \tag{8.3}$$

where v is the BK signal, t is the time, and $\mathcal{V}(t/T)$ is the universal scaling function. Fig. 8.7 shows the test of the universal scaling laws reported in Eq. 8.3 for a partially crystallized magnetic ribbon. The most remarkable feature is the time asymmetry of

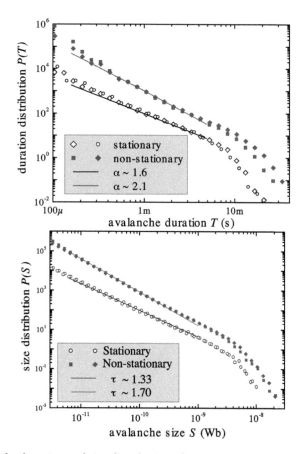

Fig. 8.5 Avalanche duration and size distributions for a stationary and non-stationary signal. Critical exponents α and τ of Eqs. 8.1–8.2 are fitted in the linear part of the plots. Two values of the applied field frequency (10 mHz and 20 mHz) are reported, showing that the distributions do not depend on the driving rate. Source: Durin and Zapperi 2006.

the shapes, similar to what's found in Spasojevic *et al.* (1996) where the average of avalanches is made without scaling.

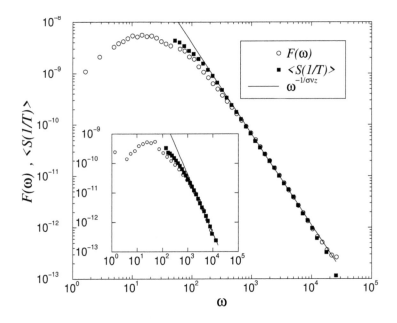

Fig. 8.6 Comparison of the power spectrum $F(\omega)$ with the average size $\langle S \rangle$ as a function of the inverse of duration T, for the amorphous material and the polycrystalline SiFe ribbon (7.8 %, inset) of Fig. 8.2. The theoretical prediction of Kuntz and Sethna (2000) is also shown, with $\gamma_{sT} = 1/\sigma\nu z$ equal to 1.77 and 2. Source: Durin and Zapperi (2002), Fig. 1.

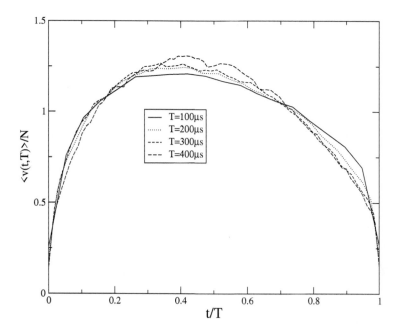

Fig. 8.7 The average pulse shape obtained from Barkhausen noise measurements in a partially crystallized $Fe_{64}Co_{21}B_{15}$ ribbon. The shapes for different durations T are normalized and rescale quite well, apart from a small systematic variation. The shapes are not completely symmetric. Source: Zapperi *et al.* (2005*a*)

8.2   Energetics of ferromagnetic materials

Ferromagnetic materials are usually thought of as an ensemble of interacting spins, as in the case of the Ising model discussed in Chapter 1. It is thus tempting to start from these classes of models to explain the experimental results reported in section 8.1. Indeed the non-equilibrium RFIM described in Chapter 3 was originally introduced to account for the experimental measurements of the Barkhausen effect, relating the observed scaling behavior to a disorder-induced critical point. The RFIM model, however, neglects several important ingredients present in ferromagnetic materials, such as dipolar interactions, and a quantitative experimental verification of this universality class is problematic. In fact, while a disorder induced non-equilibrium phase transition in the hysteresis loop has been observed experimentally in Co-CoO films (Berger *et al.*, 2000) and Cu-Al-Mn alloys (Marcos *et al.*, 2003), the measured values of the critical exponents are as yet unexplained theoretically. In addition, it was not possible to directly measure avalanches in these experiments. That said, the broad universality class of the RFIM could find some unexpected experimental applications. As in equilibrium critical phenomena a universality class is characterized by a set of symmetries and conservation laws, so s_i is not necessarily interpreted as a magnetic spin. In this respect it has been suggested that capillary condensation of helium in a porous medium is well described by the RFIM (see section 10.3).

In section 3.4 we listed a series of variants of the RFIM emphasizing different aspects of ferromagnetic materials. Here we take a step backwards and discuss more generally the properties of ferromagnetic materials. We will then come to a more suitable description of the experimental results. It is often customary to describe the local magnetization with a vectorial field $\mathbf{M} = M_s\mathbf{m}$, where M_s is the saturation magnetization and $|m| = 1$. The total energy is written as a sum of different terms depending on $\mathbf{m}$

$$E = E_{ex} + E_H + E_{dm} + E_{an} + E_{dis}, \qquad (8.4)$$

where E_{ex} represents exchange interactions, E_H the energy due to external field, E_{dm} the dipolar energy, E_{an} the anisotropy, and E_{dis} the disorder.

The most important energetic contribution comes from exchange interactions, which are typically short-ranged and tend to align spins as in the Ising model. In the continuum limit the exchange energy can be written as

$$E_{ex} = A \int d^3r \sum_{\alpha=1}^{3} (\nabla m_\alpha(\mathbf{r}))^2, \qquad (8.5)$$

where A is the exchange coupling, which can be derived from the microscopic interaction from a coarse grained procedure. The coupling with the external magnetic field $\mathbf{H}$ is given by

$$E_H = -\frac{\mu_0}{8\pi} \int d^3r \mathbf{M} \cdot \mathbf{H}. \qquad (8.6)$$

The two terms above represent a continuum version of a Heisenberg Hamiltonian and reduce to the Ising model in the limit of infinite anisotropy, when the magnetization is aligned on a specific "easy" axis.

In a real ferromagnet, we should also consider the energy associated, the field generated by the local magnetization because of dipolar interactions. To this end we should solve the Maxwell equation, which in the magnetostatic limit reduces to

$$\nabla \cdot \mathbf{B} = 0 \qquad \nabla \times \mathbf{B} = 0. \tag{8.7}$$

Thus from the relation $\mathbf{B} = \mu_0(\mathbf{H}_{dm} + \mathbf{M})$ we obtain

$$\nabla \cdot \mathbf{H}_{dm} = -\nabla \cdot \mathbf{M}. \tag{8.8}$$

Noticing that the irrotational condition on the field (Eq. 8.7) implies that it can be written as the gradient of a potential, we obtain an intuitive picture defining a *magnetic charge* density $\rho \equiv -\nabla \cdot \mathbf{M}$. The problem now reduces to an electrostatic one and the energy is readily computed as

$$E_{dm} = -\frac{\mu_0}{8\pi} \int \frac{d^3r \, d^3r' \rho(\mathbf{r})\rho(\mathbf{r}')}{|\mathbf{r} - \mathbf{r}'|}, \tag{8.9}$$

which can also be written explicitly as

$$E_{dm} = -\frac{\mu_0}{8\pi} \int d^3r d^3r' \sum_{\alpha,\beta=1}^{3} \left(\frac{\delta_{\alpha\beta}}{|\mathbf{r} - \mathbf{r}'|^3} - \frac{3(r_\alpha - r'_\alpha)(r_\beta - r'_\beta)}{|\mathbf{r} - \mathbf{r}'|^5} \right) M_\alpha(\mathbf{r})M_\beta(\mathbf{r}'). \tag{8.10}$$

Magnetic charges can also arise at the surface of the sample if the magnetization has a discontinuity. In general a surface separating two regions of magnetizations $\mathbf{M}_1$ and $\mathbf{M}_2$ is associated with a surface charge density

$$\sigma = \hat{n} \cdot (\mathbf{M}_1 - \mathbf{M}_2), \tag{8.11}$$

where $\hat{n}$ is the vector normal to the surface. At the boundary of the sample, where the magnetization varies abruptly from M_s to zero, the charge density is given by $M_s \cos\theta$, where θ is the angle between the magnetization direction and $\hat{n}$. These surface charges are responsible for the *demagnetizing field* $\mathbf{H}_{dm}$, which for a uniformly magnetized ellipsoid is constant and proportional to the magnetization vector

$$\mathbf{H}_{dm} = -k\mathbf{M}, \tag{8.12}$$

where k is a geometry-dependent demagnetizing factor . In more complicated geometries $\mathbf{H}_{dm}$ is not strictly constant but always opposes the magnetization (Cerruti et al., 2009). The corresponding energy is simply obtained by replacing $\mathbf{H}$ in Eq. 8.6 with $\mathbf{H} + \mathbf{H}_{dm}/2$.

Magnetization in a ferromagnetic material typically has preferential directions corresponding to the crystallographic axis of the material. This general observation is quantified by the energy of magnetocrystalline anisotropy

$$E_{an} = \int d^3r \sum_{\alpha,\beta} K_{\alpha\beta}m_\alpha m_\beta, \tag{8.13}$$

where m_α is the α component of the vector $\mathbf{m}$ and $K_{\alpha\beta}$ is a symmetric tensor, describing the anisotropy of the material. In the simplest case of uniaxial anisotropy, Eq. 8.13 reduces to

$$E_{an} = \int d^3r K_0 (\mathbf{m} \cdot \hat{e})^2 = \int d^3r K_0 M^2 \sin^2 \phi, \tag{8.14}$$

where ϕ is the angle between the easy axis $\hat{e}$ and the magnetization vector and K_0 is the uniaxial anisotropy constant. The variations of the magnetization inside a ferromagnetic sample can cause deformation in the lattice structure, a phenomenon known as magnetostriction. Conversely, when an external mechanical stress is applied to the sample, the magnetic structure can, in principle, be modified. To describe this effect, it is useful to introduce the magnetoelastic energy, which in the most general form can be written as

$$E_{an} = \int d^3r \sum_{\alpha,\beta,\gamma,\delta} \lambda_{\alpha\beta\gamma\delta} \sigma_{\alpha\beta} m_\gamma m_\delta, \tag{8.15}$$

where $\sigma_{\alpha\beta}$ is the stress tensor and $\lambda_{\alpha\beta\gamma\delta}$ is the magnetoelastic tensor. For a crystal with isotropic magnetostriction, under a uniaxial stress σ, the anisotropy energy takes the simple form of Eq. 8.14 with K_0 replaced by $K_0 + 3/2\lambda\sigma$, where λ is the uniaxial magnetostriction constant.

In the previous discussion we have considered only an homogenous system, in which the interactions are globally defined and do not depend on position. In general, however, different sources of inhomogeneities are found in virtually all ferromagnetic materials. The presence of structural disorder is essential to understand the fluctuations in the Barkhausen noise, which would be strongly suppressed in a perfectly ordered system. The nature of the disorder can be inferred from the microscopic structure of the material under study.

We can thus distinguish several contributions to the magnetic free energy due to the disorder: In crystalline materials disorder is due to the presence of vacancies, dislocations, and non-magnetic impurities. In polycrystalline materials we should add to these defects the presence of grain boundaries and variations of the anisotropy axis in different grains. Finally, in amorphous alloys disorder is primarily due to internal stresses and the random arrangement of the atoms. It is important to notice that in the following we will consider the disorder as quenched (or "frozen"): it does not evolve on the timescales of the magnetization processes under study. It is not always simple to quantify the energetic contributions of the different sources of disorder, but we can highlight here some important effects.

The presence of randomly distributed non-magnetic inclusions give rise to a magnetostatic contribution due to the magnetic charges that form at the boundaries of the inclusions (Néel, 1946), and to a fluctuating exchange coupling. The typical fluctuations of the coupling depend on the volume fraction v of the non-magnetic inclusions: The typical strength of the (exchange or dipolar) coupling $g(\mathbf{r})$ will be of the order of $\bar{g} \simeq (1-v)g_0$ and the fluctuations $\langle (g - \langle g \rangle)^2 \rangle \simeq v g_0^2$, where g_0 is the coupling without impurities. This type of disorder is conventionally called random-bond and should also be present in amorphous alloys, due to the random arrangement of the atoms. In polycrystalline samples, each grain has a different crystalline anisotropy. In particular, the direction of the anisotropy axes will fluctuate in space, and in the simple case of uniaxial anisotropies the anisotropy energy will be given by $E_{an} = \int d^3 r K (\mathbf{M} \cdot \hat{e}(\mathbf{r}))^2$, where $\hat{e}$ is a random function of position reflecting the grain structure of the material under study. In magnetostrictive samples, internal stresses play a similar role to anisotropies. In particular, a random distribution of internal stresses produces a random energy of the type of Eq. 8.14 with a random anisotropy constant $K(\mathbf{r}) \propto K_0 + 3/2\lambda\sigma(\mathbf{r})$.

In summary, we can collect all the terms and write Eq. 8.4 explicitly as

$$E = \sum_{\alpha=1}^{3} \int d^3 r [A(\nabla m_\alpha)^2 + K(m_\alpha e_\alpha)^2 - (H_\alpha + (1/2)H_{dem}^{(\alpha)})M_s m_\alpha], \qquad (8.16)$$

A, K and e_α in principle depend on position and $H_{dem}^{(\alpha)}$ is the component α of the demagnetizing field, including surface and bulk contributions. The energy function reported in Eq. 8.16 can in principle be used to compute the equilibrium properties of ferromagnetic materials. This problem is in general very complex and we do not attempt to address it here. In fact, to describe the Barkhausen effect we are interested in the evolution of the magnetization in response of an increasing external field. The equation of motion for the magnetization can readily be obtained from Eq. 8.16

$$\frac{\partial \mathbf{m}}{\partial t} = \Gamma_m \mathbf{m} \times \mathbf{H}_{eff} \qquad (8.17)$$

where Γ_m is the charge to mass ratio and $\mathbf{H}_{eff} \equiv -\delta E/\delta \mathbf{M}$. Eq. 8.17 does not include the dissipation mechanism and would predict an indefinite precession of the magnetization vector. This problem can be overcome by introducing phenomenological laws for the dissipation, as in the Landau–Lifshitz–Gilbert equation that can be integrated for different microstructures and boundary conditions. It is tempting to use this micromagnetic approach to describe collective effects arising in the Barkhausen effect from the first principles. To this end, it would be necessary to study three-dimensional models with long-range interactions, a task that lies beyond the present computational capabilities.

8.3 Domain walls

In order to minimize the exchange energy, a ferromagnetic material at sufficiently low temperature and zero field would display a uniform magnetization, oriented along an easy axis to lower the magnetocrystalline anisotropy energy. In most cases, however, this arrangement is hindered by a very high cost in magnetostatic energy, due to the discontinuities of the magnetization at the sample edges. Thus it is energetically more favorable to break up the sample into domains of opposite magnetization. When this occurs, the magnetization process occurs predominantly by domain wall motion, which is then analyzed to model the statistical properties of the Barkhausen noise (Urbach et al., 1995a; Narayan, 1996; Cizeau et al., 1997; Zapperi et al., 1998; Bahiana et al., 1999; de Queiroz and Bahiana, 2001).

Domain walls can be of several types, depending on the way the magnetization vector rotates from one domain to anther. A very common example is provided by

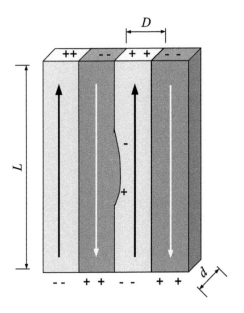

Fig. 8.8 The idealized domain structure of soft ferromagnetic materials. The discontinuities of the magnetization at the boundary of the sample induces a demagnetizing field that can be computed using fictitious magnetic charges. These also appear the in the presence of local deformations of the domain walls .

Bloch walls, in which the magnetization is always parallel to the domain wall surface. In this way, no magnetic charges are produced and the wall structure results from the competition between anisotropy and exchange energies. A domain wall of width δ_W has an (uniaxial) anisotropy energy cost for unit surface $E_{an} \sim K\delta_W$ and an exchange cost $E_{ex} \sim A/\delta_W$. Minimizing the total wall energy $\gamma_W \equiv E_{an} + E_{ex}$ with respect to δ_W, we obtain $\delta_W \sim \sqrt{A/K}$ and $\gamma_W \sim \sqrt{AK_0}$. This result can be refined by minimizing the complete micromagnetic equation, but the general dependence on K and A is correct. Similarly, the spacing D between domain walls in a sample can then be obtained by considering the competition between the domain wall energy γ_W and the magnetostatic energy. In the simple case of a set of parallel domain walls (see Fig. 8.8), for a sample of length L and thickness d, the creation of $n = L/D$ domain walls costs an energy $E_W = n\gamma_W$ but reduces the magnetostatic energy by $E_{dm} - \mu_0 M_s^2 D/d$. By minimizing the sum of these two contributions we obtain $D = \sqrt{\gamma_W d/(\mu_0 M_s^2 L)}$.

In fact, extended parallel domains are commonly observed in soft magnetic ribbons, the typical materials in which the Barkhausen noise measurements were discussed in section 8.1. In the absence of disorder, the domain wall would be flat, but in general the wall will bend to accommodate pinning forces. Having this in mind, we consider a flexible 180° domain wall separating two regions with opposite magnetization directed along the z axis. If the surface has no overhangs, we can describe the position of the domain wall by a function $u(\mathbf{r}, t)$ of space and time. The energy associated with such a deformation field can be computed by expanding the micromagnetic free energy discussed in section 8.2, then we can split the energy into the sum of different contributions due to exchange and magnetocrystalline interactions, magnetostatic and dipolar fields, and disorder (Zapperi *et al.*, 1998).

The contribution from the magnetostatic energy, if the external field **H** is applied along the z axis, is given by

$$E_m = -2\mu_0(H + H_{dm})M_s \int d^2r\, u(\mathbf{r}, t), \tag{8.18}$$

where we have included the demagnetizing field H_{dm} which is in general a complicated function of $u(\mathbf{r}, t)$, depending also on the sample shape. In the simplest approximation, however, the intensity of the demagnetizing field will be proportional to the total magnetization. Considering the field constant through the sample, it can thus be written as

$$H_{dm} = -\frac{kM_s}{V} \int d^2r\, u(\mathbf{r}, t), \tag{8.19}$$

where the demagnetizing factor k takes into account the geometry of the domain structure and the shape of the sample, and V is a volume factor.

As discussed above, the domain wall energy due is given for the unit surface by γ_W and hence the total energy is

$$E_{dw} = \gamma_W \int d^2r\sqrt{1 + |\nabla u(\mathbf{r}, t)|^2}, \tag{8.20}$$

where we have multiplied γ_W by the wall surface area. Expanding this term for small gradients we obtain

$$E_{dw} = \gamma_w S_{dw} + \frac{\gamma_W}{2} \int d^2r |\nabla u(\mathbf{r}, t)|^2, \tag{8.21}$$

where S_{dw} is the area of the undeformed wall. Thus the domain wall energy represents an elastic interaction that tends to keep the domain wall flat.

An additional elastic interaction is due to dipolar forces, since the local distortions of the domain wall are associated with discontinuities in the normal component of the magnetization. As discussed in section 8.2, this effect can be treated by introducing magnetic charges on the domain wall surface (see Eq. 8.11). The surface charge density is zero when the magnetization is parallel to the wall and for small distortions it can be expressed as

$$\sigma(\mathbf{r}) = 2M_s \cos\theta \simeq 2M_s \frac{\partial u(\mathbf{r}, t)}{\partial x}, \tag{8.22}$$

where θ is the local angle between the vector normal to the surface and the magnetization. The energy associated with this distribution of charges is given by

$$E_d = \int d^2r d^2r' \frac{\mu_0 M_s^2}{2\pi|\mathbf{r} - \mathbf{r}'|} \frac{\partial u(\mathbf{r}, t)}{\partial z} \frac{\partial u(\mathbf{r}', t)}{\partial z'}, \tag{8.23}$$

which, integrating by part, can also be written as

$$E_d = \int d^2r d^2r' K(\mathbf{r} - \mathbf{r}')(u(\mathbf{r}, t) - u(\mathbf{r}', t))^2, \qquad (8.24)$$

where the non-local kernel is given by

$$K(\mathbf{r} - \mathbf{r}') = \frac{\mu_0 M_s^2}{2\pi |\mathbf{r} - \mathbf{r}'|^3} \left(1 - \frac{3(z - z')^2}{|\mathbf{r} - \mathbf{r}'|^2} \right). \qquad (8.25)$$

The interaction is long range and anisotropic, as can be seen by considering the Fourier transform

$$K(p, q) = \frac{\mu_0 M_s^2}{4\pi^2} \frac{p^2}{\sqrt{p^2 + q^2}}, \qquad (8.26)$$

where p and q are the two components of the Fourier vector along z and y. In the preceding derivation we have implicitly assumed an infinitely strong anisotropy, so that the magnetization never deviates from the easy axis.

The disorder present in the material in the form of non-magnetic impurities, lattice dislocations, or residual stresses is responsible for the deformation and the pinning of the domain wall. In general, disorder can be modelled by introducing a random potential $V(\mathbf{r}, h)$, whose derivative gives the local pinning field $\eta(\mathbf{r}, h)$ acting on the surface. In the particular case of point-like defects, such as non-magnetic impurities, the random force is given by

$$\eta(\mathbf{r}, h) = -\sum_i f_p(\mathbf{r} - \mathbf{r}_i, h - h_i), \qquad (8.27)$$

where (r_i, h_i) are the coordinates of the pinning centers and $f_p(x)$ is the individual pinning force, which typically has a range comparable with the domain wall width $\delta_W \simeq \sqrt{K/A}$. After coarse-graining at a scale larger than the typical distance between the pinning centers, this disorder can be replaced by a Gaussian random noise with correlations

$$\langle \eta(\mathbf{r}, h)\eta(\mathbf{r}', h') \rangle = \delta^2(\mathbf{r} - \mathbf{r}')R(h - h'), \qquad (8.28)$$

where $R(x)$ decays very rapidly for large values of the argument. It has been shown that the particular form of $R(x)$ (i.e. due to random-bond or random-field types of disorder) does not have a relevant effect on the scaling laws associated with the domain wall dynamics (Narayan and Fisher, 1993). Another possible source of pinning is due to local variations of the domain wall energy γ_W (Néel, 1946), due for instance, to random anisotropies. Thus the domain wall energy becomes a function of position $\gamma(\mathbf{r}, h)$ which we can expand about its average $\gamma_W + \eta(\mathbf{r}, h)$. Introducing this expression in Eq. 8.21, we obtain to lowest order, an additional random term $\eta(\mathbf{r}, h)$, whose distribution and correlations can be directly related to the random anisotropies.

8.4   Domain wall dynamics

8.4.1   Depinning and avalanches

In metallic materials, the motion of the domain walls is strongly overdamped, because of eddy current dissipation. In the quasistatic limit, eddy currents yield an equation of motion for the wall given by

$$\Gamma \frac{\partial u(\mathbf{r}, t)}{\partial t} = -\frac{\delta E(\{u(\mathbf{r}, t)\})}{\delta u(\mathbf{r}, t)}, \tag{8.29}$$

where $E(\{u(\mathbf{r}, t)\})$ is the total energy functional, derived in the previous section, and Γ is an effective viscosity. We have neglected here thermal effects, since experiments suggest that these are not relevant for the Barkhausen effect in bulk three dimensional samples (Urbach *et al.*, 1995*b*). This could be different in thin films (Lemerle *et al.*, 1998) where thermally activated motion can be described by adding an extra noise term to the equation of motion.

Collecting all the energetic contributions, we obtain the equation of motion for the domain wall (Zapperi *et al.*, 1998). In order to avoid a cumbersome notation, all the unnecessary factors can be absorbed in the definitions of the parameters. The equation then becomes

$$\frac{\partial u(\mathbf{r}, t)}{\partial t} = H - \bar{k}\tilde{u} + \gamma_W \nabla^2 u(\mathbf{r}, t) +$$

$$\int d^2 r' K(\mathbf{r} - \mathbf{r}')(u(\mathbf{r}') - u(\mathbf{r})) + \eta(\mathbf{r}, h), \tag{8.30}$$

where the dipolar kernel K is reported in Eq. 8.25, the effective demagnetizing factor is given by $\bar{k} \equiv 4\mu_0 k M_s^2/V$, and $\tilde{u} \equiv \int d^2 r' u(\mathbf{r}', t)$. Owing to the fact that the demagnetizing field term is just an approximation, the dependence of k on the sample shape and size can be quite complex. Variants of Eq. (8.25) have been extensively studied in the past (Néel 1946; Hilzinger 1976; 1977; Enomoto 1994*a*; 1994*b*, 1997; Urbach *et al.* 1995*a*; Narayan 1996; Cizeau *et al.* 1997; Zapperi *et al.* 1998; Bahiana *et al.* 1999; de Queiroz and Bahiana 2001).

When the demagnetizing factor k is negligible, as for instance in a frame geometry, Eq. 8.30 displays a depinning transition as a function of the applied field H and the domain wall moves only if the applied field overcomes a critical field H_c. The critical behavior associated with the depinning transition has been studied using renormalization group methods (Nattermann *et al.*, 1992; Narayan and Fisher, 1993; Ertas and Kardar, 1994; Leschhorn *et al.*, 1997; Chauve *et al.*, 2001). As discussed in Chapter 4, if we consider an interface whose interaction kernel in momentum space scales as $K(q) = A_K|q|^\mu$, the upper critical dimension is given by $d_c = 2\mu$ and the values of the exponents depend on μ. Thus, in general we would expect mean-field behavior, although there are situations in which the dipolar coupling could be neglected with respect to the domain wall tension γ_W where one effectively could observe the critical behavior associated with $\mu = 2$. The renormalization group analysis allows for a determination of the critical exponents in the framework of $\epsilon = d_c - d$ expansion. As discussed above, for $\mu = 1$ we expect mean-field results $\tau = 3/2$ and $\alpha = 2$, while for $\mu = 2$ to first order in ϵ one obtains $\tau = 5/4$ and $\alpha = 11/7$. A two loop expansion has been carried out, allowing for an estimate to order $O(\epsilon^2)$ of the exponents ($\tau = 1.24$ and $\alpha = 1.51$) (Chauve *et al.*, 2001). Finally, numerical simulations yield $\tau = 1.27$ and $\alpha = 1.5$ for $\mu = 2$ (see Table. 4.3).

The discussion above applies strictly to the case $k = 0$ and scaling requires that H is close to H_c. If one instead ramps the field slowly, sampling the avalanche distribution over all the values of the field, the result will be different. We need to integrate the distribution over H (Zapperi *et al.*, 1998) as discussed in section 4.5.2 obtaining $\tau_{int} = \tau + \sigma$. Using the values of the exponents discussed in Chapter 4, we obtain $\tau_{int} = 1.72$ for $\mu = 2$ and $\tau_{int} = 2$ for $\mu = 1$. A similar discussion can be repeated for avalanche duration, yielding $\alpha_{int} = \alpha + \Delta$. The numerical result is $\alpha_{int} = 2.3$ for $\mu = 2$ and $\alpha_{int} = 3$ for $\mu = 1$. These values can be used to account for the experiments performed in non-stationary conditions (Durin and Zapperi, 2006).

In a typical Barkhausen experiment $k > 0$ and the effective field acting on the domain wall is given by $H_{eff} = ct - k\bar{h}$. Thus, when $H_{eff} < H_c$ the domain wall is pinned and the effective field increases. As soon as $H_{eff} > H_c$ the domain wall acquires a finite velocity and $\bar{h}$ increases. As a consequence, the effective field will be reduced until the domain wall stops. This process keeps the domain wall close to depinning transition, but criticality is only reached in the limit $c \to 0$ and $k \to 0$. It is easy to recognize here the general mechanism of self-organized criticality, as discussed already in previous chapters. The demagnetizing factor k represents a restoring force hindering the propagation of Barkhausen avalanches. Thus when $k > 0$, we expect to find a finite avalanche size cutoff S_0. The scaling of the cutoff was discussed in section 4.5.2 and the corresponding results can be applied here directly. Using Eqs. 4.44-4.62, we obtain $1/\sigma_k = 0.71$ and $\Delta_k = 0.34$ for $\mu = 2$, and $1/\sigma_k = 2/3$ and $\Delta_k = 1/3$ in mean-field theory. The numerical values provide a convincing explanation of the experimental results reported in Fig. 8.4.

8.4.2   Mean-field theory and the ABBM model

The mean-field theory, which provides a good qualitative description of the depinning transition and describes quantitatively the data for $d = d_c$ (apart from logarithmic corrections), is obtained by discretizing the equation of motion, coupling all the sites with the average domain wall position $\bar{u}$ (Fisher, 1985). The dynamics of such an infinite range model is described by

$$\frac{du_i}{dt} = ct - k\bar{u} + J(\bar{u} - u_i) + \eta_i(u), \tag{8.31}$$

where J is an effective coupling and $i = 1 \ldots N$. Summing over i both sides of Eq. (8.31), one obtains an equation for the total magnetization $m = \sum u_i$:

$$\frac{dm}{dt} = \tilde{c}t - km + \sum_{i=1}^{N} \eta_i(u). \tag{8.32}$$

We could now interpret $\sum_i \eta_i$ as an effective pinning $W(m)$ and determine its correlations.

When the domain wall moves between two pinned configurations, W changes as

$$W(m') - W(m) = \sum_{i=1}^{n} \Delta\eta_i, \tag{8.33}$$

where the sum is restricted to the n sites that have effectively moved (i.e. their disorder is changed). The total number of such sites scales as $n \sim l^d$ and in mean-field theory is proportional to the avalanche size $S = |m' - m|$ (since $S \sim l^{d+\zeta}$ and $\zeta = 0$). Assuming that the $\Delta\eta_i$ are uncorrelated and have random signs, we obtain a Brownian effective pinning field

$$\langle |W(m') - W(m)|^2 \rangle = D|m' - m|, \tag{8.34}$$

where D quantifies the fluctuation in W.

In this way the mean-field theory turns out to be equivalent to a phenomenological model for the Barkhausen noise introduced by Bertotti (1986, 1987), who included a Brownian pinning field in a random energy model. The model, widely known as ABBM (Alessandro *et al.* 1990*a*; 1990*b*), yields a quite accurate description of the Barkhausen noise statistics (Colaiori, 2008). In the ABBM model the magnetization evolves according to an overdamped equation of motion

$$\frac{dm}{dt} = ct - km + W(m), \tag{8.35}$$

where the external field increases at constant rate $H = ct$, k is the demagnetizing factor, and the damping coefficient has been set to unity, rescaling the time units. Thus the Brownian pinning field, introduced phenomenologically in the ABBM model, is an effective description of the disorder resulting from the collective motion of a flexible domain wall. As a consequence of this mapping, we expect that the infinite-range model will display the same frequency dependence of the exponents as in the ABBM model as confirmed by simulations.

The main predictions of the ABBM model can be obtained by deriving Eq. (8.35) with respect to time and defining $v \equiv dm/dt$:

$$\frac{dv}{dt} = c - kv + vf(m), \tag{8.36}$$

where $f(m) \equiv dW/dm$ is an uncorrelated random field with variance D. Expressing Eq. 8.36 as a function of v and m only

$$\frac{dv}{dm} = \frac{c}{v} - k + f(m), \tag{8.37}$$

we obtain a Langevin equation for a random walk in a confining potential, given by $U(v) = kv - c\log(v)$. Asymptotically, the statistics of v is ruled by the Boltzmann distribution

$$P(v, m \rightarrow \infty) \sim \exp(-U(v)/D) = v^{c/D} \exp(-kv/D). \tag{8.38}$$

The distribution in the time domain is obtained by a simple transformation and it is given by (Alessandro *et al.*, 1990*a*; Colaiori, 2008)

$$P(v) \equiv P(v, t \rightarrow \infty) = \frac{k^{c/D} v^{c/D-1} \exp(-kv/D)}{D^{c/D} \Gamma(c/D)}. \tag{8.39}$$

A consequence of Eq. 8.39 is that the domain wall average velocity is given by $\langle v \rangle = c/k$. For $c/D < 1$ the velocity distribution in Eq. 8.39 is a power law with an upper cutoff that diverges as $k \to 0$. In this regime, the domain wall moves in avalanches whose sizes and durations are also distributed as power laws. For $c/D > 1$ the motion is smoother with fluctuations that decrease as c/D increases.

The avalanche size distribution equals the distribution of first return times of a random walk in the confining potential $U(v)$. In the limit $k \to 0$, one can compute the distribution of the first return times of a random walk in a logarithmic potential, which implies that the avalanche exponents depend on c and are given by (Colaiori, 2008)

$$\tau = 3/2 - c/2D \qquad \alpha = 2 - c/D. \tag{8.40}$$

See also Durin *et al.* (1995*a*, 1995*b*) for alternative but not rigorous derivations of these results.

The scaling of the cutoff of the avalanche distributions in the ABBM model can be obtained similarly in the limit $c \to 0$, solving for the first return probability of a biased random walk, and is given by

$$s_0 \sim k^{-2}. \tag{8.41}$$

Using similar arguments one can also show that the cutoff of avalanche durations scales as $T_0 \sim k^{-1}$ (Zapperi *et al.*, 1998).

The domain wall model discussed above describes rather well the avalanche statistics and the power spectrum of Barkhausen noise in soft magnetic materials. Despite this fact, the pulse shape in the model proves to be perfectly symmetric, in marked contrast with experiments that systematically display an asymmetric shape. To explain this discrepancy, we notice that Eq. 8.29 is derived assuming that at each time the work done by the effective field is compensated by the energy dissipated by eddy currents, which is estimated in the quasistatic approximation (Alessandro *et al.*, 1990*a*). A more detailed analysis of eddy current dissipation, including dynamic effects (Bishop, 1980; Zapperi *et al.*, 2005*a*), leads to the identification of a frequency-dependent effective mass, which turns out to be negative in the entire spectrum. This negative inertial effect can also be formulated by adding a non-local damping term in the equation of motion for the wall

$$\Gamma \frac{\partial u(\mathbf{r}, t)}{\partial t} + \frac{\Gamma_0}{\tau} \int^t e^{-(t-t')/\tau} \frac{\partial u(\mathbf{r}, t')}{\partial t'} = -\frac{\delta E(\{u(\mathbf{r}, t)\})}{\delta u(\mathbf{r}, t)}, \tag{8.42}$$

where Γ and Γ_0 are coefficients of the same order of magnitude and τ is the largest eddy current relaxation time. Numerically integrating Eq. 8.42 in the mean-field limit (ABBM), one obtains pulse shapes bearing a remarkable similarity to the experimental ones. From this study one can conclude that an asymmetric pulse shape is an effect of inertia: A leftward asymmetry corresponds to a negative mass and conversely a rightward asymmetry to a positive mass.

The role of eddy currents in determining the asymmetry of the Barkhausen noise pulse shape was then confirmed in an experimental study of Barkhausen noise in permalloy thin films (Papanikolaou *et al.*, 2011). A careful study of the average avalanche shapes led to symmetric shapes undistorted by eddy currents that are suppressed in thin samples. Furthermore, the shapes of the pulses were found to be in excellent quantitative agreement with the predictions of the ABBM model (Papaniko-laou *et al.*, 2011).

8.5   Magnetic avalanches in thin films

In thin film, inductive Barkhausen measurements are far less accurate than in bulk materials, due to the small signal-to-noise ratio. Indeed, when the thickness is small the unavoidable air flux becomes important. Some attempts were still made in the past using a fine compensation of air flux by an additional coil (Wiegman 1977, 1979; Wiegman and Stege 1978) which aided in observing the noise in permalloy films down to a thickness of 40 nm. This good result is, however, quite isolated and is probably due to the particularly high mobility of the domain walls in that material. In most cases the only valid alternative to inductive methods is provided by the magneto optical Kerr effect (MOKE), which is based on the change of polarization of the light reflected by a magnetic material. This technique is extensively used to investigate the domain structure of most magnetic materials (Hubert and Schäfer, 1998). There are two variations of the basic MOKE setup applied to the investigation of Barkhausen noise and in particular to the statistical distributions. The first was introduced in (Puppin *et al.*, 2000) adding an optical stage to tune the laser spot size onto the film, with values ranging from 20 μm up to 700 μm, allowing one to estimate the distribution of avalanche sizes in various thin films Puppin (2000). The second solution makes use of an advanced video processing technique to directly observe the domains and their motions (Kim *et al.* 2003*a*, 2003*b*). Despite these promising results, the investigation of Barkhausen avalanches in magnetic thin films is still at an early stage.

While bulk three-dimensional systems often display a relatively simple magnetic structure with parallel domain walls , thin films show richer and often more com-plicated DW patterns. The film thickness plays a fundamental role because of the increasing importance of stray fields in the direction perpendicular to the sample (Néel, 1955). In addition to usual Bloch walls, we observe as well the complicated patterns of 90° Nèel walls, cross-tie walls, and zigzag walls, to name just a few. A complete description of the existing configurations together with many experimental images, can be found in the excellent book by Hubert and Schäfer (1998). Thus, in order to attempt a theoretical description of Barkhausen experiments in thin film, it is necessary first to clarify the the domain structure under study.

The inductive results obtained in Wiegman (1977, 1979); Wiegman and Stege (1978) allow the observation of the crossover from three to two dimensions. The scaling exponents τ and α do not show an appreciable thickness dependence above 100 nm, in correspondence to the region where complex cross-tie walls are normally observed in this material. While τ and α fluctuate strongly for these small thicknesses, the power spectrum exponent $1/\sigma\nu z$ displays a sharp transition from 2 to 1.5. We notice also that the scaling exponents have the same values observed in polycrystalline ribbons $(\tau \sim 1.5, \alpha \sim 2, 1/\sigma\nu z \sim 2)$ for thicknesses down to 100 nm, indicating that a crossover between two and three-dimensional behavior is probably taking place below this scale.

Inductive measurements are problematic at low thickness, and the quantitative results reported so far in the literature are mostly obtained with magneto-optical methods. In (Puppin, 2000), the exponent τ is estimated in a Fe film of 90 nm using the MOKE setup and considering different laser spot sizes from 20 to 700 μm. A single distribution $P(\Delta M)$ is obtained, properly rescaling the size of the avalanche ΔM and the distribution P with the size of the spot. The estimated exponent is $\tau = 1.1 \pm 0.05$. Kim *et al.* (2003*a*) report an estimate of the exponent τ in Co films, using a MOKE setup which allows a direct visualization of the domain walls and obtaining the size distribution measuring the area swept by a wall between two consecutive images (see Fig. 8.9 for a similar experiment reported in Magni *et al.* (2009)). In this experiment, the field is kept at a constant value, very close to the coercive field, and the Barkhausen activity is due to thermally activated avalanches. The resulting power-law distribution has an exponent $\tau = 1.33$, averaged over four different thicknesses, from 5 to 50 μm. In a later paper (Ryu *et al.*, 2007), the same group analyzed with the same technique the domain wall dynamics in MgAs thin films, a material with a Curie temperature slightly above room temperature. By increasing the temperature, it was shown that the domain-wall morphology is crossing over from a zigzag to a rough structure. The crossover is also reflected in the avalanche exponent, which is changing from $\tau = 1.33$ for the zigzag wall to $\tau = 1.1$ for the rough wall. The main effect of temperature in this material is to change the saturation magnetization, effectively tuning the dipolar interaction strength. Finally, methodological problems with the extraction of avalanche sizes from images have been discussed in Magni *et al.* (2009). Boundary effects can bias the distribution estimate because avalanches can be truncated by the edges. The most reliable estimate of the exponent is obtained by including only avalanches that do not touch the edges. The exponent measured in Magni *et al.* (2009) for permalloy thin film is $\tau = 1.15$. From these results it appears that the statistical analysis of Barkhausen noise in thin films is still at a preliminary stage, but the results reached so far are very promising.

Other techniques, such as magnetic force microscopy (Schwarz *et al.*, 2004) or Hall sensors (Damento and Demer, 1997), have also emerged as powerful tools to study Barkhausen noise at the nanoscale. Schwarz *et al.* (2004) measured Barkhausen avalanches in thin films with labyrinthine domains which can be reproduced using a phase field model of a ferromagnet with out-of-plane magnetization (Benassi and Zapperi, 2011). Another set of experiments revealed Barkhausen jumps due to a domain wall depinning from the atomic Peierls potential (Novoselov *et al.*, 2003; Christian *et al.*, 2006). In this experiment, the authors were able to measure a power-law distribution with an exponent $\tau = 1.1$ (Christian *et al.*, 2006). It would be interesting to have independent estimates of Barkhausen noise exponents using different techniques on the same sample. This would greatly improve the reliability of the measurements.

To understand the experimental results obtained in thin films requires an analysis of two-dimensional models. This is not as straightforward as it might look, since several new features arise when going from three to two dimensions. The demagnetizing field vanishes for very thin samples and the dipolar interaction displays a different behavior depending on whether the magnetization lies in or out of the film plane. In addition, the domain structure and the single domain walls are often more complicated than in the bulk. Despite these difficulties, it is still instructive to extend the three dimensional to two dimensions. The equation of motion for a single domain wall in two dimensions with in-plane magnetization is similar to the one in three dimensions

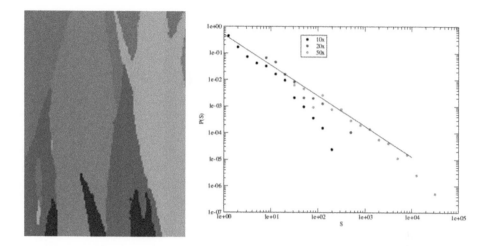

Fig. 8.9 Jump sequences on a 170 nm permalloy thin film. Magnetic field direction is vertical. The grayscale indicates the switching time (left). The associated avalanche size distribution for different magnifications (right). Source: Magni *et al.* (2009).

(e. g. Eq. 8.30), but the demagnetizing factor k is virtually zero. In Fourier space, the dipolar kernel (Eq. 8.25) does not scale as q but as $q^2 \log(aq)$ where a is a small-scale cutoff (Nattermann, 1983). Apart from the logarithmic correction, the dipolar term is similar to the domain wall energy and thus, in principle, should not significantly affect the universality class of the depinning transition. The problem then reduces to the depinning transition of a linear interface with short-range interactions (see Table 4.3). While these values are close to some of the experimental results reported in the literature, the agreement needs to be supported by a more careful investigation of the domain structure. It would then be necessary to understand Barkhausen noise due to more complicated domain walls . In this respect, Cerruti and Zapperi (2006) report a study of the dynamics of a model for a magnetically charged "zigzag" domain wall. The avalanche distributions yield exponents $\tau = 1.33$ and $\alpha = 1.5$. These values are compatible with the experiments reported in Ryu *et al.* (2007) in the low-temperature limit, where the dipolar interaction becomes relevant because the magnetization is perpendicular to the domain wall, leading to the zigzag instability.

In conclusion, the investigation of Barkhausen noise in thin films represents the real frontier of the field. Promising experimental techniques are currently being developed and explored. We can expect in the near future that these will yield interesting results that will pose new exciting challenges to theorists.

9

Vortices in superconductors

The transport properties of vortices in type II superconductors represent a challenging theoretical problem in condensed matter. Due to the interplay between vortex repulsion and pinning, the magnetic flux forms a steady profile, known as the Bean state. Experiments suggested analogies with SOC sandpile models since magnetic flux enters in avalanches in the material. In several cases, these avalanches are power-law distributed and this phenomenon can be accounted for by models resembling the SOC sandpile. In this chapter, we first discuss briefly the basic properties of type-II superconductors, from the microscopic interactions between individual vortices to the phenomenological macroscopic behavior of flux profiles. Next, we provide an account of the main experimental observations of vortex avalanches. The efforts made in the past to understand the observed avalanches from the motion of individual vortices is summarized, and we finally discuss the path from microscopic to macroscopic models.

9.1 Magnetic properties of type-II superconductors

When a superconductor is placed in a small magnetic field H, magnetic flux is expelled: the so-called Meissner effect. There is, however, a temperature-dependent critical field $H_c(T)$ above which the flux starts to penetrate in the sample and superconductivity is lost. The behavior we have just described is characteristic of type-I superconducting materials, but in several cases the situation is more complex than that. Type-II superconductors display an intermediate phase for $H_{c1} < H < H_{c2}$ where flux penetrates only partially into the sample. This mixed state could naively be considered as a domain state in a ferromagnet, with alternating normal and superconducting regions. If we estimate the domain wall energy, however, we would find a *negative* value, meaning that the total length of these interfaces would proliferate (de Gennes, 1966). In fact, as was first shown by Abrikosov, magnetic flux penetrates in the sample in the form of vortex lines, each carrying a flux quantum $\Phi_0 = h/2e$, where e is the electron charge and h is the Plank constant (Abrikosov, 2004).

The magnetic field inside a supeconductor obeys the London–London equation, which, in the presence of a vortex line directed along $\hat{z}$ in the origin, reads as

$$\lambda^2 \nabla \times \nabla \times \mathbf{B} + \mathbf{B} = \hat{z}\Phi_0 \delta^2(r), \tag{9.1}$$

where $\lambda = 4n_s e^2/m^* c^2$ is the penetration length, expressed in terms of the superfluid concentration n_s and the effective mass m^*. Solving Eq. (9.1), we obtain the field around a vortex line

$$\mathbf{B} = \hat{z}\frac{\Phi_0}{2\pi\lambda^2}K_0(|\mathbf{r}|/\lambda), \tag{9.2}$$

where $K_0(x)$ is a Bessel function decaying exponentially for $x \gg 1$. From Ampere law $\mathbf{J} = c/(4\pi)\nabla \times \mathbf{B}$, we can also compute the supercurrent density that surrounds the vortex line. Notice that Eq. 9.2 displays a logarithmic divergence at the origin, that is in practice cut off by the coherence length ξ. Due to the induced magnetic field s, a pair of vortex lines repels each other with an interaction energy

$$E_{12}(\mathbf{r}) = [\Phi_0^2/(8\pi^2\lambda^2)]K_0(|\mathbf{r}|/\lambda). \tag{9.3}$$

This mutual repulsion leads to the formation of a triangular vortex lattice, as first shown by Abrikosov. Vortex lines also interact with impurities in the host material, such as vacancies, dislocations, or grain boundaries, which provide effective pinning centers, distorting the perfect crystal order of the Abrikosov lattice and hindering vortex motion. There is a very extensive literature devoted to the investigation of *vortex phases*, resulting from the competition of vortex interactions, thermal fluctuations, and quenched disorder (for detailed reviews see Blatter *et al.* (1994, Brandt (1995)). High-temperature superconductors display a wide variety of vortex arrangements: crystals, polycrystals, glasses, and liquids.

Here we leave aside the important issue of equilibrium phases and concentrate instead to the non-equilibrium dynamics of vortices which is related to the avalanche phenomena under investigation throughout this work. When the external magnetic field is raised beyond H_{c1}, vortices are nucleated at the boundary of the sample and are slowly pushed inside it. Due to a combination of pinning impurities and vortex mutual repulsion, a flux density profile is typically formed across the sample. Bean was the first to describe phenomenologically this state postulating that the local current density in the superconductor is either zero or equal to a *critical current* J_c (Bean, 1962). If we consider a superconducting slab of length L in the x direction and infinite in the other directions, from Ampere law we obtain $\mathbf{B} = B(x)\hat{z}$, where $B(x)$ is a piecewise linear function decreasing from the sample edge (see Fig. 9.1). As the external field is changed, the flux profile is shifted upwards and downwards accordingly. The Bean model is very successful in describing the average magnetic properties of type-II superconductors, but does not take into account fluctuations. In fact, as we will discuss in the next section, flux enter in the superconductor through intermittent avalanches and we thus need a description that goes beyond mean-field quantities.

9.2 Experimental measurements of vortex avalanches

Experiments on vortex avalanches have been nicely reviewed in Altshuler and Johansen (2004). Here, we discuss the main point and update the review with recent developments reported in the literature.

The first hints that vortex motion in the Bean state came from inductive measurements performed with an experimental setup of the type employed for the Barkhausen effect. A pickup coil is wound across a superconducting sample placed into a solenoid. The voltage in the coil measures the variations in the magnetic flux across the sample. In 1968 it was shown that the flux noise in PbIn alloys displayed a broad distribution, indicating vortex avalanches (Heiden and Rochlin, 1968). In those times, before scaling and critical phenomena came to light, not much attention was given to power laws and

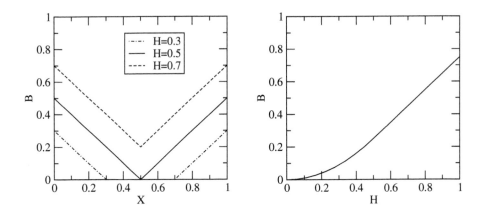

Fig. 9.1 A schematic picture of the Bean model. The left panel shows the evolution of the flux profile in the sample as the magnetic field H is ramped. The profile is a piecewise linear function with a slope equal to $\pm 4\pi J_c/c$. The right panel reports the corresponding B–H curve. The initial part of the curve is quadratic up to the critical state when the flux profiles meet. From this point on the curve is linear.

the authors focused only on the exponential cutoff of the distributions. It was shown in this way that the typical flux change $\Delta\Phi$ was a function of the applied field showing a maximum around $B \simeq 500$Oe. If we replot the data of Heiden and Rochlin in log–log scale, we see that indeed the exponential cutoff is preceded by a power-law regime (see Fig. 9.2). This observation was confirmed with greater accuracy in Field *et al.* (1995) using a slightly different experimental setup: A pickup coil was placed inside a hollow superconduting cylinder of a NbTi alloy. In this way, the outer radius of the sample was placed in a field **H**, the inner radius was at zero field, and the pickup coil recorded the number of flux lines pushed outside the sample (see Fig. 9.3). The distribution of flux avalanches was found to be a power law over more than one decade, with an exponent and cutoff that depend on the applied field value (the exponents were found in the range $1.4 < \tau < 2.2$). We notice that these results could be biased by the cutoff and one could interpret the data considering a single exponent and a field-dependent scaling function.

Inductive methods are not the only option to investigate flux changes in supercon-ductors; different methods have also been used and have improved our understanding of this phenomenon. Aegerter (1998) used a magnetometer to extract the flux avalanches from the magnetization changes in a relaxation experiment in BSCCO crystals. He found a power law distribution over less than two decades, with an exponent $\tau \simeq 2$. Arrays of micro-hall probes help to obtain a precise measure of the local flux inside a superconductor. This setup was used by Behnia *et al.* (2000) to analyze the spa-tiotemporal dynamics of vortices by recording local flux changes in granular Nb films. Flux changes in a single probe were found to obey a power law with exponent $\tau \simeq 2$ over a limited regime, followed by an exponential cutoff. It was stated in Behnia *et al.* (2000) that the cutoff value was too low to be attributed to finite size effects. Sim-ilar experiments were performed by Altshuler *et al.* (2004) in Nb films, who found

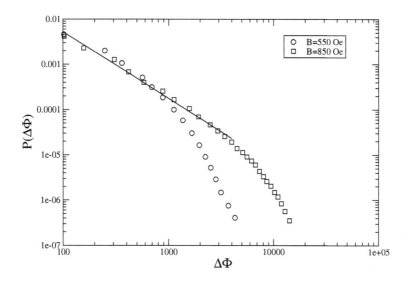

Fig. 9.2 The avalanche distribution measured for two different values of the applied field, plotted in log-log scale. The avalanche size is measured by the flux change $\Delta\Phi$ entering the PbIn sample. The line is a power law with exponent $\tau = 1.5$. Source: Data from Heiden and Rochlin (1968).

a power-law distribution of flux changes over two decades with an exponent $\tau \simeq 3$. An important point discussed in Altshuler *et al.* (2004) was the influence of the flux landscape, revealed by magneto-optical methods, and the noise measured by the Hall probe.

While the measurements discussed above clearly establish the presence of widely distributed vortex avalanche, they do not provide a coherent picture of the scaling properties. The power-law regime is often limited, resulting in a wide range for the measured exponents. In addition, it is difficult to clearly visualize the nature of the microscopic processes involved in vortex avalanches. A fundamental step forward in this direction was made by the careful experiments performed by the group of Wijngaarden. Using subsequent sets of magneto-optical images, it was possible to follow in detail the formation of the Bean profile and the formation of flux avalanches as the external magnetic field was slowly ramped. This obtains a coherent scaling picture for vortex avalanches, linking their statistical properties to the geometry of the Bean profile. In particular, impurities lead to fluctuations on the flux profile and roughening of the flux front.

If the sample edge is across the y direction, the magnetic field inside the sample can be expressed as $\mathbf{B} = \hat{z}B(x, y)$. The set of zero intersection of $B(x, y)$ defines the flux front $h(y)$[1] and the cuts of $B(x, y)$ at constant x are used to study the flux profile $\rho(x)$. The experiments of Wijngaarden provide a direct image of $B(x, y)$ for subsequent times, allowing for a direct analysis for the spatiotemporal dynamics of

---

[1] Notice that in practice one considers the intersection $B(h(y), y) = B_0$, where B_0 is higher than the background noise.

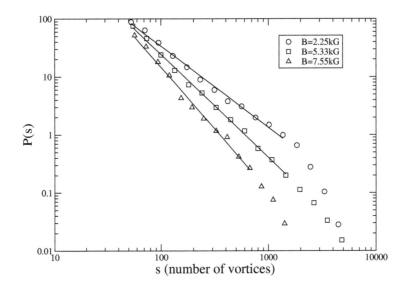

Fig. 9.3 Avalanche distribution associated with the voltage noise recorded for a cylinder of NiTi alloy. Here, the size s represents the number of vortices leaving the superconductors. The scaling exponent depends on the external field at which the noise is collected. Source: Data from Field *et al.* (1995).

the Bean profile. Depending on the strength of disorder, which has been controlled for instance by changing the concentration of H impurities in NbH_x or by tuning the degree of epitaxy in YBCO, one observes a crossover from self-affine to self-similar flux fronts, reminiscent of the behavior discussed in section 3.5 for domain walls in the RFIM. For weak disorder regimes, both the flux front and the flux profiles are self-affine and one can apply the standard tools used in surface growth phenomena. In particular, the roughness exponent is found to be $\zeta = 0.67$ for flux fronts in YBCO (Surdeanu *et al.*, 1999) and Nb (Vlasko-Vlasov *et al.*, 2004) and $\zeta = 0.75$ for flux profiles in YBCO (Welling *et al.*, 2004). The dynamic roughness exponents are given by $\beta = 0.6$ (Surdeanu *et al.*, 1999) for fronts and $\beta = 0.7$ for profiles (Welling *et al.*, 2004). We notice here that (Welling *et al.*, 2004) reports instead lower values for the roughness of the profiles that were attributed to a spurious correction due to the background noise.

The magneto-optical images of flux penetration are very useful to determine the distribution of vortex avalanches. Subtracting two subsequent images, one can then define the avalanche size as

$$s \equiv \int dx dy \Delta B(x, y), \tag{9.4}$$

where the integral is either taken above the whole measured area (Aegerter *et al.*, 2004) or restricted to each avalanche, in the case in which several separated events are distinguishable in the entire region (see Fig. 9.4 and Welling *et al.* (2005)). By recording data on regions of different size L, it is possible to test for finite-size scaling in the avalanche size distribution

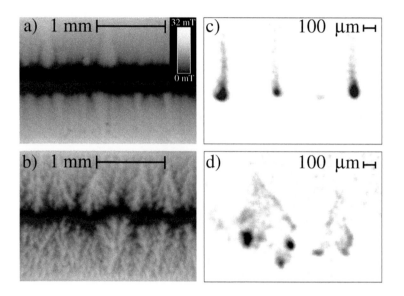

Fig. 9.4 The magnetic flux landscape in the center of a pure Nb film (a) and with added hydrogen impurities (b). Magnetic flux avalanches are obtained by the subtraction of two consecutive images for the two cases (c)-(d). Source: Reprinted from Welling *et al.* (2005).

$$P(s, L) = s^{-\tau} g(s/L^D), \tag{9.5}$$

with $\tau = 1.29$ and $D = 1.89$ in YBCO films (Aegerter *et al.*, 2004). In (Welling *et al.*, 2005) a similar study was performed on NbH$_x$ films with varying H-concentration. For pure Nb the fronts are relatively smooth and become fractal as the H-concentration is increased. The avalanches in the first case do not obey finite size scaling, while they do so when the fronts are fractal, defining the exponents $\tau = 1.1$ and $D = 2.25$ (see Fig. 9.5 and Welling *et al.* (2005)).

9.3 Vortex dynamics simulations

In an infinitely long sample, rigid flux lines can be modeled as a set of interacting particles performing an overdamped motion in a random pinning landscape (Pla and Nori 1991; Barford *et al.* 1993; Olson *et al.* 1997*a*, 1997*b*; Monier and Fruchter 2000). The equation of motion for each flux line i can be written as

$$\Gamma \mathbf{v}_i = \sum_j \mathbf{f}_{vv}(\mathbf{r}_i - \mathbf{r}_j) + \sum_p \mathbf{f}_{vp}[(\mathbf{R}_p - \mathbf{r}_i)/l] + \eta(\mathbf{r}_i, t), \tag{9.6}$$

where the effective viscosity can be expressed in terms of material parameters $\Gamma = \Phi_0 H_{c2}/\rho_n c^2$. Here, Φ_0 is the magnetic quantum flux, c is the speed of light, ρ_n is the resistivity of the normal phase and H_{c2} is the upper critical field. The first term on the right-hand side represents the vortex–vortex interaction obtained from Eq. 9.3

$$\mathbf{f}_{vv}(\mathbf{r}) \equiv -\frac{dE_{12}}{d\mathbf{r}} = [\Phi_0^2/(8\pi\lambda^3)]K_1(|\mathbf{r}|/\lambda)\hat{r}, \tag{9.7}$$

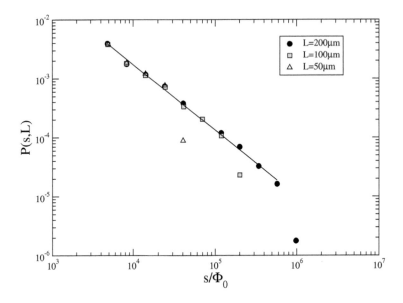

Fig. 9.5 The distribution of avalanche sizes for a Nb film with hydrogen impurities recorded for different window sizes L. The data are well fit with a power law with exponent $\tau = 1.1$ (line). Source: Data from Welling *et al.* (2005).

where the function K_1 is a Bessel function decaying exponentially for $|\mathbf{r}| > \lambda$ and λ is the London penetration length (de Gennes, 1966). The second term on the right-hand side accounts for the interaction between pinning centers, modeled as localized traps, and flux lines. Here, $\mathbf{f}_{vp}$ is the force due to a pinning center located at $\mathbf{R}_p$, l is the range of the wells (typically $l \ll \lambda$), and $p = 1, ..., N_p$ (N_p is the total number of pinning centers). Pla and Nori (1991, Barford *et al.* (1993), Olson *et al.* (1997a, 1997b) employed parabolic traps, but to to avoid discontinuities in the force one can use instead a form of the type (Monier and Fruchter, 2000) $\mathbf{f}_{vp}(\mathbf{x}) = -f_0 \mathbf{x}(|\mathbf{x}|-1)^2$, for $|\mathbf{x}| < 1$ and zero otherwise. Next, we consider an uncorrelated thermal noise term η, with zero mean and variance $\langle \eta^2 \rangle = k_b T / \Gamma$, although most simulations are restricted to the case $T = 0$ (see Monier and Fruchter (2000) for the implementation of thermal noise in MD simulations).

If we are interested in the response to an external current J, we should add to Eq. 9.6 the Lorenz force $\mathbf{F}_L = \Phi_0 \mathbf{J} \times \hat{z}$. Simulations in this case are done by imposing periodic boundary conditions and studying the depinning transition of the system. The problem of a moving vortex lattice translates to a particular elastic manifold moving through a disordered substrate where the depinning force corresponds to the critical current. In these conditions vortex avalanches are expected as we ramp the current towards the critical value and are indeed observed in simulations. To compare the model with experiments discussed in section 9.2, however, we should bear in mind that the magnetic field , rather than the current, is driving the activity. To simulate the increase of the magnetic field , one should implement a mechanism for the injection of vortices through the boundary of the sample. In earlier simulations this was done by injecting vortices in a pin-free region at the boundary of the sample and then letting the system relax.

A more sophisticated treatment of vortex injection was introduced in Shumway and Satpathy (1997), combining a typical vortex dynamics simulation scheme with a Monte Carlo method for vortex nucleation. The Gibbs potential associated with N vortices of coordinates $\mathbf{r}_i$ can be written as

$$G = \sum_{ij} E_{12}(\mathbf{r}_i - \mathbf{r}_j) + \sum_i \tau_i(x_i) - \frac{H}{4\pi} \sum_i \phi_i(x_i). \tag{9.8}$$

If we consider a semi-infinite system, bounded by the $y = 0$ line, we must add to the vortex–vortex interaction a term accounting for the interaction between each vortex and the image of the others (Shumway and Satpathy, 1997). Similarly, the term $\tau_i(x_i)$, where x_i is the distance between the vortex i and the sample surface, represents the interaction between each vortex and its own image (Shumway and Satpathy, 1997). Finally, the external magnetic field gives rise to an effective chemical potential with $\phi(x) \equiv \Phi_0(1 - \exp(-x/\lambda))$.

The dynamics of the front penetration was studied in Reichhardt *et al.* (1996, Zapperi *et al.* (2001, Moreira *et al.* (2002) using vortex dynamics simulations. The general idea is to reconstruct flux fronts and profiles from a suitable coarse-graining of the vortex arrangements (see Fig. 9.6). In addition, we have direct access to the avalanches following individual vortex motion (Olson *et al.*, 1997a). When the total number of vortices is kept constant, the average front position $\bar{h}(t)$ grows initially as $t^{1/3}$ for small times. Eventually the front position slows down and saturates to a value ξ_p which increases as the strength of disorder decreases (see Fig. 9.6). In particular, for a system with pinning center of density n and strength f_0 the pinning length scales as $\xi_p \sim (f_0 \sqrt{n})^{-1/2}$ as could be expected on the basis of collective pinning arguments. These scaling laws crucially depend on the boundary condition imposed on the simulations, as will be clarified by the coarse-grained approach discussed in the next section. The avalanche fluctuations were studied in detail in Olson *et al.* (1997a) as a function of the pinning strength and the pinning density. While the results obtain a qualitative picture of the avalanche phenomenology, it is difficult to perform a direct comparison with experiments. The main problem is that the way to quantify avalanche amplitudes in simulations, computing the duration or the total vortex displacement, does not correspond directly to the flux variations measured experimentally.

9.4 Coarse-grained models, flux diffusion, and cellular automata

In order to understand the results of vortex dynamics and link them to macroscopic constitutive laws, it is instructive to perform a coarse-graining of Eq. 9.6. The Fokker-Plank equation for the probability distribution of the flux line coordinates $P(\mathbf{r}_1, ..., \mathbf{r}_N, t)$ is given by

$$\Gamma \frac{\partial P}{\partial t} = \sum_i \nabla_i(-\mathbf{f}_i P + k_B T \nabla_i P), \tag{9.9}$$

where $\mathbf{f}_i$ is the force on the particle i given by Eq. (9.6). Next, we introduce the single particle density $\rho(\mathbf{r}, t) \equiv \langle \sum_i \delta^2(\mathbf{r} - \mathbf{r}_i) \rangle$, where the average is over the distribution

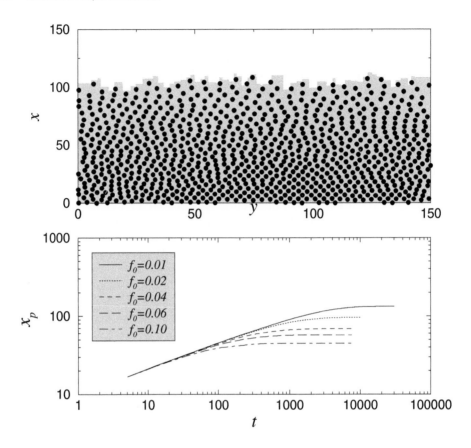

Fig. 9.6 In the upper panel, we show a typical realization of a flux front obtained from a simulation of interacting vortices in a disordered landscape (Moreira *et al.*, 2002). In the lower panel, we show the growth of the average front position as a function of time for different pinning force strengths. The front initially grows as $t^{1/3}$ and then is pinned at a distance that scales with f_0. Source: Miguel *et al.* (2003).

$P(\mathbf{r}_1, ..., \mathbf{r}_N, t)$. The evolution of ρ can be directly obtained from Eq. (9.9) and is given by

$$\Gamma \frac{\partial \rho}{\partial t} = -\nabla \cdot \left(\int d^2r' \mathbf{f}_{vv}(\mathbf{r} - \mathbf{r}') \rho^{(2)}(\mathbf{r}, \mathbf{r}', t) \right.$$

$$\left. - \sum_p \mathbf{f}_{vp}[(\mathbf{R}_p - \mathbf{r})/l] \rho(\mathbf{r}, t) \right) + k_B T \nabla^2 \rho, \tag{9.10}$$

where $\rho^{(2)}(\mathbf{r}, \mathbf{r}', t)$ is the two-point density, whose evolution depends on the three-point density and so on. The simplest truncation scheme involves the approximation $\rho^{(2)}(\mathbf{r}, \mathbf{r}', t) \simeq \rho(\mathbf{r}, t)\rho(\mathbf{r}', t)$. We then coarse-grain the equation considering length scales larger than λ. This can be done by expanding $\mathbf{f}_{vv}$ in Fourier space, keeping only the lowest-order term in $\mathbf{q}$, and retransforming back into real space. The result reads

$$\int d^2r' \mathbf{f}_{vv}(\mathbf{r} - \mathbf{r}')\rho(\mathbf{r}', t) \simeq -a\nabla\rho(\mathbf{r}, t), \tag{9.11}$$

where $a \equiv \int d^2r\mathbf{r} \cdot \mathbf{f}_{vv}(\mathbf{r})/2 = \Phi_0^2/4$.

The coarse-graining of the disorder term is more subtle: A straightforward elimination of short-wavelength modes gives rise, as in the previous case, to a random force $\mathbf{F}_c(\mathbf{r}) = -g\nabla n$, where n is the coarse-grained version of the microscopic density of pinning centers $\hat{n}(\mathbf{r}) \equiv \sum_p \delta^2(\mathbf{r} - \mathbf{R}_p)$ and $g \propto f_0$. This method, however, cannot be applied for short-range attractive pinning forces, since short wavelength modes yield a macroscopic contribution to pinning that cannot be neglected. For instance, consider the flow between two coarse-grained regions: short-range microscopic pinning forces give rise to a macroscopic force that should always oppose the motion, while the random force derived above could in principle point in the direction of the flow. In other words, $F_c(\mathbf{r})$ is a *friction* force, whose direction is always opposed to the driving force $\mathbf{F}_d$ and whose absolute value is given by $|g\nabla n|$ for $|\mathbf{F}_d| > |g\nabla n|$ and by $|\mathbf{F}_d|$ otherwise.

Collecting all the terms, we finally obtain a disordered non-linear diffusion equation for the density of flux lines:

$$\Gamma\frac{\partial\rho}{\partial t} = \nabla(a\rho\nabla\rho - \rho\mathbf{F}_c) + k_BT\nabla^2\rho. \tag{9.12}$$

The boundary conditions depend on the particular driving mode that we wish to consider, (Bryksin and Dorogovtsev 1993a, 1993b; Gilchrist and van der Beek 1994) such as:

(A) Constant total number of vortices. Experimentally this corresponds to an external control of the magnetic flux.

(B) Constant vortex concentration at the boundary. This case corresponds to an external control of the magnetic field .

(C) Total vortex number increasing at constant rate. This represents an external control of the flux rate.

(D) Boundary concentration increasing at constant rate, corresponding to a constant field rate.

As a word of caution, one should notice that boundary conditions can be more complicated in reality, due to complex surface barriers that oppose flux penetration.

For a pure system ($f_0 = 0$) at $T = 0$, Eq. (9.12) can be solved exactly using scaling methods (Bryksin and Dorogovtsev, 1993a; Bryksin and Dorogovtsev, 1993b; Gilchrist and van der Beek, 1994). In this case, the density profiles obey the equation

$$\rho(x, y, t) = t^{-\chi}\mathcal{G}(x/t^{\psi}), \tag{9.13}$$

where χ and ψ satisfy $\chi + 2\psi = 1$. For the boundary conditions considered: A) $\chi = 1/3$, $\psi = 1/3$; B) $\chi = 0$, $\psi = 1/2$; C) $\chi = -1/3$, $\psi = 2/3$; and D) $\chi = -1$, $\psi = 1$. These exponents are in perfect agreement with the results from molecular dynamics simulations reported in Moreira *et al.* (2002). For instance, the data reported in Fig. 9.6 correspond to boundary condition (A) and correctly scale with $\psi = 1/3$. The function $\mathcal{G}(u)$ also depends on the boundary condition and for the case (A) is given by $\mathcal{G}(u) =$

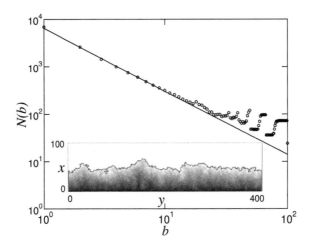

Fig. 9.7 In the inset is shown the density plot obtained simulating Eq. 9.12 for $T = 0$, $g = 5$, $s = 6.28$ and concentration at the boundary increasing at rate $h = 0.01$. In the main figure we report the box counting plot of the front, averaged over twelve realizations of the disorder. The front has fractal dimension $D_f = 4/3$ (solid line) and crosses over to $D_f = 1$ at large length scales. Source: Zapperi *et al.* (2001)

$(1 - u^2)/6$ for $u < 1$ and vanishes for $u \geq 1$. The other cases are reported in Bryksin and Dorogovtsev (1993a, 1993b), Gilchrist and van der Beek (1994).

The presence of disorder induces substantial effects on the front propagation and/or the shape of the density profiles of flux lines. Depending on the boundary conditions, it has been observed that the front is either pinned (for cases A and B) or simply slowed down (for cases C and D) (Zapperi *et al.*, 2001). Extensive numerical simulations have also been performed in Zapperi *et al.* (2001, Moreira *et al.* (2002) to show the compatibility between the molecular dynamics model with disorder and its coarse-grained representation, Eq. (9.12). Moreover, by varying the parameters of this continuum description of the front propagation, a crossover from smooth to fractal flux fronts has been detected, consistent with experimental observations. The value of the fractal dimension $D_f \simeq 1.3$ suggests that the strong disorder limit is described by a gradient percolation front (see Fig. 9.7). In the weak disorder limit, one recovers the analytical results derived in Bryksin and Dorogovtsev (1993a, 1993b), Gilchrist and van der Beek (1994).

The coarse-grained model described above is very close to the one proposed earlier by Barford (1997), derived on the basis of phenomenological considerations. In this model the equation of motion for the vortex density is given by

$$\Gamma \frac{\partial \rho}{\partial t} = -\nabla(\rho \mathbf{F}),\tag{9.14}$$

where the net force **F** has components $F_i = A(\partial_i \rho)^2 - F_i^{(p)}$. Here, $A = \Phi_0^2/(8\pi)$ and $F^{(p)}$ are random frictional forces opposing the motion. Notice that the main difference with Eq. 9.12 lies in the order of the non-linearity, but qualitatively the two equations are similar. It was noticed in Barford (1997) that such a continuum model would not yield avalanches unless one introduced a difference between static and dynamic friction forces. This means that when the flux gradient overcomes the local (static) pinning force $F^{(p)}$, one should reduce the pinning force to the dynamic value until the flow ceases, otherwise the flux density would spread like a fluid, without avalanches. The origin of this rule lies in the discrete nature of flux lines that is lost in the continuum model and should thus be enforced by hand.

Dynamic friction also can be implemented in Eq. 9.12 by considering a lower friction force $F_d < F_c$ whenever there is flow in that particular region. The nonlinear equation then becomes suitable to study avalanches. In order to reproduce the experimental conditions, one can consider a constant field boundary condition (case B) and then increase the external field ΔH in steps. The resulting magnetic behavior follows closely the predictions of the Bean model: The magnetic induction B increases quadratically with the applied field H (see Fig. 9.8(a)). This is due to the formation of a linear vortex density profile with a constant slope as in the Bean model (see Fig. 9.8(b)). While the average behavior is described in the Bean model, we also notice the presence of fluctuations. In analogy with Welling *et al.* (2005), after each field step ΔH we record the flux change $s = \int dx dy \Delta rho(x, y)$ and compute its ditribution. As shown in Fig. 9.8(c), the distribution depends strongly on the field step: A high value of ΔH induces a peak at large s which is reduced as ΔH is decreased. For sufficiently small values of ΔH the peak disappears and one can see a power-law distribution with an exponent $\tau \simeq 1.6$. Notice that a crossover between peaked distribution is observed in experiments (Welling *et al.*, 2005) changing the impurity concentration.

The continuum models described above are not very efficient at describing the internal avalanche dynamics observed in experiments, and it is more natural to resort to a cellular automaton model. This was constructed in Bassler and Paczuski (1998) defining an integer variable m_i on the sites i of a two-dimensional honeycomb lattice. A random potential V_i is also associated with each site and is chosen from a bimodal distribution with values 0 or p, occurring with probabilities $1 - q$ and q respectively. At each time step, a unit of m is transferred across each bond connecting sites i and j depending on the sign of the force $F_{ij} = V_j - V_i - 1 + (m_i - m_j) + \Delta_{ij}$, where Δ is a next-nearest-neighbor correction to the discretized gradient $m_i - m_j$. The lattice is updated in parallel until all sites are in a stable configuration. At this point units of m are injected in the system from one of the edges, while the opposite edge is kept open with $m = 0$. Periodic boundary conditions are imposed on the other direction. The model is very similar to the sandpile and in fact the result for the avalanche size distribution closely resembles the values expected for the (boundary-driven) Manna model , namely $\tau = 1.63$ and $D = 2.75$. Unfortunately these values are not in agreement with experiments, but the virtue of the model is to clearly show the analogy between SOC sandpiles and vortex dynamics in the Bean state. It would be desirable to devise a more refined cellular automaton model, taking into account the non-linearity derived in coarse-grained approaches to provide a quantitative explanation of the experiments.

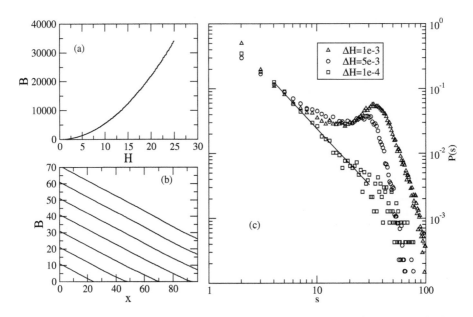

Fig. 9.8 Simulations of Eq. 9.12 using a dynamic friction value that is half of the static friction. (a) The *B–H* curve is quadratic in agreement with the Bean model and (b) is a consequence of linear flux profiles. (c) The avalanche distribution depends on the field step ΔH and is a power law in the slow driving limit. The line has a slope of $\tau = 1.6$.

10

Flow in Porous Media

The process of fluid invasion in a porous medium is often irregular and characterized by avalanches. In imbibition, one typically observes a self-affine front and avalanches are related to those observed in interface depinning. Conversely, in drainage the fluid front tears apart, giving rise to an invasion percolation process where the fluid forms a fractal interface. The crossover between these two behaviors is ruled by the wetting angle and can be described by simple models. In further applications, we discuss the capillary condensation of He in Nuclepore and air flow during the inflation of collapsed lungs. The air pressure opens the bronchial channels and alveoli in avalanches, resulting in a crackling noise.

10.1 Avalanches in the drainage and imbibition of porous media

The flow of fluids in porous media is a process with enormous practical applications for oil extraction processes, but also has important implications in geophysics. For this reason, the phenomenon has been studied in great detail through laboratory experiments, numerical simulations, and theoretical models, giving rise to an enormous body of literature that we will not attempt to review here. The interested reader can refer to extensive reviews written on the subject (Arbabi and Sahimi, 1990; Alava *et al.*, 2004). Our aim is to focus on the avalanche-like dynamics that is typically observed when one fluid displaces another inside a porous material. This behavior has been known of for a very long time, from the early observation made in 1930 by Haines of the jumps in capillary pressure in drying agricultural soils. It is only (relatively) recently, however, that this avalanche phenomenon has been studied systematically and understood via statistical physics models.

The process of fluid invasion is ruled by capillary forces that hinder the propagation of the fluid inside a pore. In the presence of two nonmiscible fluids in a capillary tube, one typically defines the *wetting* fluid as the one with a contact angle θ smaller than $\pi/2$ and the other one as the *nonwetting* fluid. The contact angle is the angle formed by the fluid interface (the meniscus) and the tube wall. The distinction is important because it affects the morphology of the invasion process which is referred to as imbibition when a wetting fluid displaces a non-wetting fluid, and as drainage in the opposite case. A very familiar example of imbibition is provided by placing a sheet of paper (the porous medium) inside a liquid container. On the other hand, drainage is observed, for instance, when air is injected into a water-filled porous rock.

In order for a nonwetting fluid to propagate into a cylindrical pore of radius r, the driving pressure should overcome the capillary pressure P_{cap}, which is linked to the fluid surface tension γ and to the contact angle θ by the relation

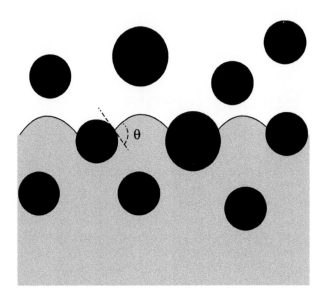

Fig. 10.1 A schematic view of a fluid invading a porous medium. At the throats the invading fluid front defines the contact angle θ with the walls.

$$P_{cap} = \frac{2\gamma \cos \theta}{r}. \tag{10.1}$$

If we schematize the porous medium by a complicated network of pores with different radii and hence different, randomly distributed, capillary pressures, we can view fluid invasion as a percolation process. This relation was confirmed experimentally by Lenormand and Zarcone (1985), who studied the structure formed by injecting non-wetting fluid (air) in a transparent etched network filled with a wetting fluid (paraffin oil). The resulting pattern was found to display a fractal structure with fractal dimension $D = 1.82$ — in good agreement with the prediction of the model of invasion percolation in two dimensions.

The experiment of Lenormand and Zarcone (1985) focused only on the patterns observed during the drainage process but provided no information on the dynamics of fluid invasion. Måløy *et al.* (1992, Furuberg *et al.* (1996) have performed a similar experiment using a randomly packed monolayer of glass beads between two transparent plastic sheets. This artificial porous medium was then filled with water and was successively, slowly drained by a pump. The water pressure was monitored at regular intervals during the drainage process, leading to a fluctuating signal. The pressure drop distribution was measured, leading to an exponential distribution. Hence, in this experiment avalanches are not power-law distributed, in contrast with invasion percolation, which would ideally predict a power law distribution. More recently, however, Xu *et al.* (2008) have reported a similar experiment in three dimensions, where a power-law avalanche distribution over a limited regime was indeed observed. In this experiment, the author observes in the confocal microscope the drying dynamics of a three dimensional porous medium obtained by a colloidal suspension. It was thus pos-

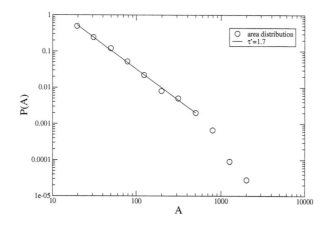

Fig. 10.2 The invaded area distribution in the drying of a three-dimensional porous medium composed by colloidal particles. The area here represents a three-dimensional cut over the three dimensional porous medium. The exponent $\tau_A = 1.7$ is obtained from a fit over the linear part. A slightly lower exponent was obtained in Xu *et al.* (2008) using a non-linear fit. Source: Data from Xu *et al.* 2008.

sible to directly visualize changes in the invaded areas as the fluid was draining. The area distribution is a power law with an exponential cutoff, as reported in Fig. 10.2. Notice that here the avalanche area is a two-dimensional cut over the three-dimensional invaded volume and, therefore, the measured exponent $\tau_A \simeq 1.7$ cannot be directly compared with a two dimensional model. The origin of the exponential cutoff in this experiment (Xu *et al.*, 2008) and in the one of Måløy *et al.* (1992, Furuberg *et al.* (1996) can be traced back to the fluid redistribution along the front during an avalanche, as we will discuss in the next section (Måløy *et al.*, 1992; Furuberg *et al.*, 1996).

The avalanche behavior of fluid invasion was also observed indirectly by measuring the electrical resistance of the porous medium as it is invaded by a conducting fluid. An experiment of this kind was performed by Thompson *et al.* (1987, Katz *et al.* (1988) injecting mercury into an evacuated (three-dimensional) sandstone sample. The electrical resistance of the sample was found to display a "devil's staircase" as a function of the fluid pressure. The resistance drop distribution follows a power law

$$P(\Delta R) \sim \Delta R^{-\tau_R}, \tag{10.2}$$

with an exponent crossing over from $\tau_R = 1.8$ to $\tau_R = 1.5$ as the ratio between gravitational and capillary pressure is decreased. Notice that while the resistance change is related to the invaded volume, the two quantities are not exactly the same since only the modification in the backbone of the percolation cluster modifies the total resistance.

The experiments discussed above all referred to a drainage process in which the front is fractal and the scaling is related to an invasion percolation process. The imbibition process was studied experimentally in Dougherty and Carle (1998) using an artificial two-dimensional porous medium similar to the one employed in Måløy *et al.* (1992), only injecting water into air. The result was a more compact flow profile with a rough front. By direct imaging, it was possible to study the distribution of avalanche sizes, defined as the volume change between two successive images. The distribution resulted in a clear exponential cutoff preceded by a possible, but not very extended, power law regime. Hence, the avalanches in imbibition are distributed exponentially as seen in drainage (Måløy *et al.*, 1992; Furuberg *et al.*, 1996) .

10.2   Modeling fluid avalanches in porous media

We discussed in the previous section the difference between drainage, characterized by a self-similar percolation type process, and imbibition where the fluid front is self-affine. Cieplak and Robbins (1988); Martys *et al.* (1991a, 1991b) have introduced a model that is able to interpolate between these two processes by varying the contact angle θ. The model is based on a two-dimensional array of disks with randomly distributed radii. For each value of the pressure difference P, the fluid interface is defined by a set of arcs joining neighboring disks. The arcs have a curvature radius r, determined by equating the pressure difference with the capillary pressure (see Eq. 10.1), and intersect at an angle θ as in Fig. 10.1. The pressure drop P is slowly increased until the interface becomes unstable. Three types of instabilities are possible: (i) bursts—it is not possible for an arc of radius r to intersect at an angle θ at both disks; (ii) touch—an arc between two disks intersecting a third disk; and (iii) overlap—two neighboring arcs overlap.

When an instability occurs, the interface moves forward, invading the porous medium by avalanche events. As the pressure is increased, avalanches become larger until a critical pressure P_c at which the invasion process never stops. By varying the contact angle, the growth morphology crosses over from compact at small value of θ to fractal for higher θ, and the two regimes are separated by a critical angle θ_c. The reader may notice the close similarity between this process and the domain wall dynamics discussed in section 3.5. In fact, one could associate fluid invasion process in terms of a disorder-induced critical behavior, although in the present case disorder is fixed. Increasing the value of θ, however, effectively reduces the interaction strength so it is analogous to reduce J in the RFIM, keeping Δ fixed.

Cieplak and Robbins (1988); Martys *et al.* (1991a, 1991b) have performed detailed finite-size scaling analysis of the growth process as a function of the pressure at constant values of θ in the two regimes. In the compact growth phase, the critical behavior is analogous to a depinning transition with the usual scaling exponents. The values of the critical exponents appear to be in good agreement with linear interface depinning values in $d = 1 + 1$ dimensions. In particular, the correlation length exponent is found to equal $\nu = 1.3$ and the avalanche size distribution is $\tau = 1.125$ (see Fig. 10.4) (Martys *et al.*, 1991b)—in close agreement with the expected results discussed in Chapter 4. Notice, however, that the (local) roughness exponent is reported to be $\zeta_{loc} = 0.85$ instead of the expected $\zeta_{loc} = 1$, a discrepancy that could be due to numerical problems. The simulations also indicate that the self-similar growth phase falls instead in the class of invasion percolation and one can relate the scaling to the usual percolation exponents. In particular, the avalanche size distribution is a power law with an exponent $\tau = 1.3$ (see Fig. 10.4). This exponent can be related to the correlation exponent ν, the fractal dimension of the invaded area D_f, and to the fractal dimension of the front D_e by the scaling relation $\tau = 1 - D_e/D_f - 1/(D_f\nu)$, derived in Martys *et al.* (1991*b*).

While the model above appears to be quite successful in reproducing the morphology of the fluid invasion process as a function of the contact angle, it also finds a series of scaling laws for the avalanche behavior that are not observed in real imbibition and drainage processes where the avalanche distribution decays exponentially. The origin of this discrepancy was clarified in Furuberg *et al.* (1996), where it was noticed that during a fluid avalanche the pressure drop P across the interface is not constant as in the model of Cieplak and Robbins (1988); Martys *et al.* (1991*a*, 1991*b*). When the fluid advances, the liquid has to be redistributed along the front, decreasing the capillary pressure. The change in capillary pressure due to an avalanche of size s can then be estimated as $\Delta P \simeq s/(\kappa n_f)$, where $\kappa = dv/dP_{cap}$ is the volume capacity of a pore, determining the rate of change in the amount of fluid volume v in a pore due to a change in capillary pressure, and n_f is the front length n_f. This observation has then been cast into a model considering a modified invasion percolation process in which random capillary pressure thresholds are assigned to the sites of a lattice (Furuberg *et al.*, 1996). The capillary pressure evolves dynamically as

$$P(t) = \frac{(V - N)}{\kappa n_f}, \tag{10.3}$$

where $V \propto t$ is the injected fluid volume and N is the invaded volume. The reader can see here once again the classical mechanism of SOC models with a restoring force coupled to the control parameter hindering the dynamics. Hence, as soon as the capillary pressure reaches the critical value P_c the invaded volume N grows rapidly, reducing P below P_c. The strength of this effective restoring force $\epsilon \equiv 1/(\kappa n_f)$ will then determine the cutoff of the avalanche distribution. The value of ϵ can be estimated from experimental data and it is found to be large enough to yield an exponential avalanche distribution, in good agreement with experiments. Notice that while the invasion percolation model of Furuberg *et al.* (1996) strictly speaking applies to fluid drainage, the same mechanism is present in imbibition and can account for the observed exponential avalanche distribution.

Finally, we would like to resume here the discussion about the resistance jumps observed during mercury injection in a porous medium. A change in the invaded volume does not necessarily correspond to a resistance drop, hence the resistance drop distribution is in general different from the avalanche size distribution. Katz *et al.* (1988) have modeled the phenomenon by a random resistor network, similar to the one discussed in Chapter 5. Pores are associated with the bonds of a cubic lattice and acquire a non-vanishing resistance when the pore is invaded by a percolation process. One then follows the evolution of the total resistance as the fraction of occupied volume increases. The resistance jump distribution is found to be a power-law distribution with exponent $\tau_R = 1.54$ at the lower bound of the experimentally measured exponent (Thompson *et al.*, 1987). In order to justify the experiments, it was proposed to include the effect of gravity transforming the invasion model into a gradient percolation process. The results appear to support this assumption, leading to a higher value of τ_R in the presence of a gradient. The precise relation between τ_R and other

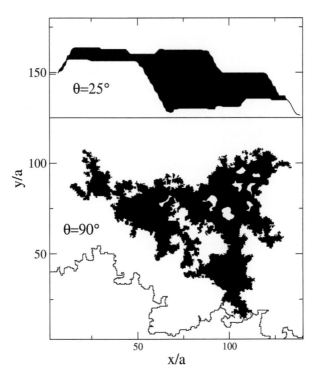

Fig. 10.3 The morphology of invaded area in the model of Martys *et al.*. Changing the contact angle θ, the morphology of the avalanches (depicted in black) crosses over from compact to fractal. Source: Data from Martys *et al.* (1991*b*).

percolation exponents was then clarified in Batrouni *et al.* (1988, Roux *et al.* (1988) through scaling arguments, leading to

$$\tau_R = 1 + \frac{d\nu}{\nu + t}. \tag{10.4}$$

Inserting percolation exponents in this relation, one obtains $\tau_R = 1.67$ in $d = 2$ and $\tau_R = 1.6$ in $d = 3$.

10.3 Capillary condensation in porous media

In a porous medium, due to the presence of capillary forces, the gas–liquid transition may occur at a chemical potential lower than the bulk value: a phenomenon known as capillary condensation. While this process can be understood theoretically for a single pore, real porous media are strongly heterogeneous with a distribution of pore lengths and sizes. Disorder in this context leads to an irregular condensation process associated with avalanche events and hysteresis.

Lilly *et al.*; (1993, 1996, 2001, 2002); Wootters and Hallock (2003) have conducted a series of experiments of capillary condensation of superfluid helium in Nuclepore, a polycarbonate material with a network of intersecting cylindrical pores with nearly

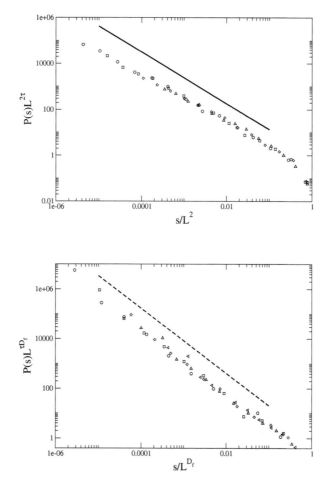

Fig. 10.4 The distribution of avalanche sizes in the fluid invasion model of Martys *et al.*. The distribution corresponding to different lattice sizes L are collapsed. (a) In the self-affine regime with $\theta = 25°$ a power law with exponent $\tau = 1.125$ is plotted as a solid line. The avalanches are compact and are thus rescaled by L^2. (b) In the percolation regime with $\theta = 179°$ a power law with a percolation exponent $\tau = 17/13 \simeq 1.3$ is plotted as a dashed line. the avalanches are fractal with fractal dimension $D_f = 1.88$. Source: Data from Martys *et al.* (1991*b*).

constant radii ($r = 200$ nm in Lilly *et al.* (1993)). In a typical experiment, a chamber in contact with the Nuclepore film is slowly filled with ^{4}He gas at constant temperature (typically $T < T_\lambda = 2.17$ K), leading to a slow increase in the chemical potential μ. When the chemical potential is large enough the gas starts to condense in the Nuclepore. The pore-filling process is monitored by measuring the capacitance of the system, thanks to the dielectric nature of liquid helium. As shown in Fig. 10.5, the condensation/evaporation process is hysteretic: The adsorption is gradual and smooth, while the drainage is more abrupt and occurs at a lower chemical potential. The difference between adsorption and desorption may just stem from the nature of the instability involved in the filling and emptying of a single pore. A description of the hysteresis process in terms of single pore events naturally leads to a Preisach-type model (Lilly *et al.*, 1993). This model, originally introduced for ferromagnetic hysteresis, considers a set of units characterized by a pair of randomly distributed chemical potentials $\mu_e < \mu_f$ for the filling and emptying of a pore. By a suitable choice of the distribution of μ_e and μ_f, one can try to reproduce the hysteresis curve in the inset of Fig. 10.5. This approach, however, fails to describe accurately the properties of the minor hysteresis loops, indicating that the adsorption/desorption process is correlated (Lilly *et al.*, 1993). In particular, the desorption process is characterized by avalanches of pore drainage events. The avalanche size distribution was measured in Lilly *et al.* (1996) and found to be a power law with exponent $\tau \simeq 2$ when averaged over all events in the desorption process. The exponent was found to be lower (i.e. $\tau = 1.5$) when sampled only over the initial part of the process (see Fig. 10.5). This is not surprising since the avalanche signal is clearly non-stationary and higher exponents are generally expected when we sample over the entire process.

The reader may have perceived the analogy between the present capillary condensation phenomenon and the disorder-induced phase transtion in the hysteresis loop of the RFIM (see Chapter 3). Indeed a lattice gas generalization of the RFIM adapted to describe the capillary condensation hysteresis was studied on a qualitative level in Guyer and McCall (1996). A similar disordered lattice gas model was analyzed in more detail in Kierlik *et al.* (2001); Detcheverry *et al.* (2003, 2005); Rosinberg *et al.* (2003). The porous medium is schematized by a three dimensional lattice with fluid occupation variables $\tau_i = 0, 1$ and quenched random variables η_i denoting the presence or absence of a pore on site i. The Hamiltonian then reads

$$\mathcal{H} = -w_{ff} \sum_{\langle ij \rangle} \eta_i \eta_j \tau_i \tau_j - w_{mf} \sum_{\langle ij \rangle} \eta_i (1 - \eta_j) \tau_i + \eta_j (1 - \eta_i) \tau_j - \mu \sum_i \tau_i, \qquad (10.5)$$

where the double sums run over nearest-neighbor pairs. For a given realization of the quenched disorder η_i, the density $\rho_i = \langle \eta_i \tau_i \rangle$ in each pore is obtained by solving the mean-field equations for the free energy:

$$\rho_i = \frac{\eta_i}{1 + \exp{-\beta(\mu + \sum_{j/i} \rho_j (w_{ff} \rho_i + w_{mf}(1 - \eta_i)))}}, \qquad (10.6)$$

where $\beta = 1/k_B T$ and the sum runs over the nearest neighbors of site i. To take into account the role of the surface, a slab with $\eta_i = 0$ is introduced at the sample edge, providing a vapour (i.e. $\tau_i = 0$) reservoir.

The behavior of the hysteresis loop of the lattice gas model depends on the distribution of quenched variables β_i, typically chosen as an uncorrelated set of particles with density ρ_m, and on the interactions ratio $y \equiv w_{ff}/w_{mff}$ (Rosinberg *et al.*, 2003). In close analogy with the RFIM, by keeping ρ_m and T constant, one can observe a transition from a discontinuous hysteresis loop at small y to a smooth loop at large y. Notice, however, that the shape of the loop branches and the nature of the transition itself is different along the adsorption/desorption curves.

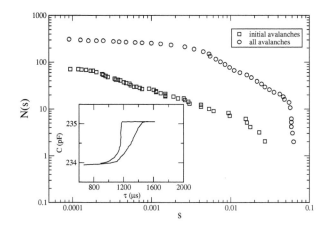

Fig. 10.5 The cumulative number of avalanches in capillary condensation in Nuclepore. The two distributions are sampled over the initial stage (yielding $\tau \simeq 1.5$) or over all the avalanches (yielding $\tau \simeq 2$). Inset: A typical hysteresis loop measured in Nuclepore, recording the capicitance during capillary condensation of He. Source: Data from Lilly and Hallock (2001)

Desorption is influenced by the vapour reservoir at the surface and proceeds as an interface growth process: For weak disorder (or low y) the fluid inside the medium has a compact structure and the fluid front undergoes a depinning transition at $\mu = \mu_c$. The jump in the hysteresis loop thus corresponds to the depinning point and disappears for strong disorder (or high y) where the front is self-similar and the process resembles invasion percolation. Thus, the scaling behavior in the desorption branch closely follows the behavior of domain wall dynamics in the RFIM discussed in section 3.5 (Ji and Robbins 1991, 1992 Koiller *et al.*1992, 2000). On the other hand, the absorption process is not influenced by the presence of a vapour surface. The jump in the hysteresis loop occurs when a network of liquid regions forms in the bulk of the sample. The transition to a smooth loop is then of the same type as the one observed in the RFIM (see section 3.3) (Sethna *et al.*, 1993). This is confirmed by the finite size scaling analysis of the model, yielding critical exponents in close agreement with those expected for the RFIM.

10.4 Avalanches in lung inflation

An interesting example of avalanche-like fluid flow in a "porous" medium is observed in the inflation of degased lungs. Air flows from the trachea down to the airways and finally to the alveoli. In normal breathing conditions, the airways and the alveoli are open and the air flow is regular. When the lung is degased, however, a part of the airways is blocked and a pressure drop is required to open them. This closure condition is encountered in some pathological situations but could also occur during surgery. We also mention the fact that all mammals must inflate the lungs when born. Due to the very heterogeneous and hierarchical nature of the lung, the opening pressures of the closed airways is typically not uniform and gives rise to an avalanche-like response.

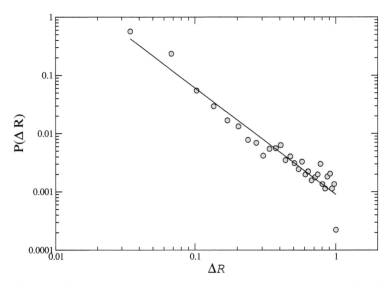

Fig. 10.6 The distribution of airway flow resistance changes in the lung. The fit yields an exponent $\tau_R = 1.8$. Source: Data from (Suki *et al.*, 1994).

The lung inflation process was studied in great detail by Suki *et al.* (1994), both experimentally and numerically. In a typical experiment, a degased lung was placed in a pressure chamber and slowly inflated. The fluctuations in the lung opening process were then recorded, measuring the changes in the terminal airway flow resistance (Suki *et al.*, 1994) and AE, by placing a microphone in the trachea (Alencar *et al.*, 1999). These methods yield power-law distributions of avalanche-like events. In particular, the flow resistance changes ΔR were found to be distributed as (see Fig. 10.6) (Suki *et al.*, 1994)

$$P(\Delta R) \sim \Delta R^{-\tau_R}, \tag{10.7}$$

with $\tau_R \simeq 1.8$. The amplitude of acoustic activity A is distributed as (Alencar *et al.*, 1999)

$$P(A) \sim A^{-\tau_A}, \tag{10.8}$$

with $\tau_A = 2.77$.

A simple model to account for the avalanche behavior of lung inflation schematizes the lung as a branching tree of airways with random opening pressures p_{ji}, where j is the generation number and $i = 1, \ldots 2^j$ (see Fig. 10.7) (Barabasi *et al.*, 1996; Sujeer *et al.*, 1997). An airway opens when the external pressure P, that is ramped slowly, exceeds the threshold p_{ji}. One can readily see that this model is equivalent to a percolation process in a Cayley tree, and hence can be solved exactly, yielding mean-field exponents. In particular, the distribution of volumes changes at constant pressure P, following a power-law distribution with exponent $\tau = 3/2$. Since the pressure is ramped slowly, we should integrate the distribution obtaining an effective exponent $\tau_{int} = \tau + \sigma = 2$. The volume distribution is difficult to measure experimentally, but indirect evidence suggests that this scaling is indeed correct (Suki *et al.*, 2000). Simulations of the model also yield the distribution of resistance changes that are in

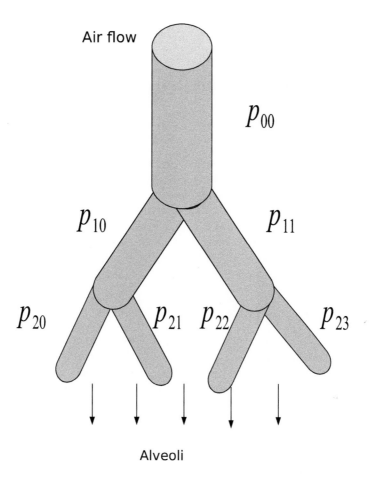

Fig. 10.7 A schematic view of the lung. The air flows from the trachea down to the alveoli through a branched tree. One can model the opening process by assigning random thresholds p_{ij} to each airway.

good agreement with the experiments (Suki *et al.*, 1994). Finally, the model can also be used to compute the pressure volume curve for the first inflation of degased lungs. Considering that the alveolar volume is approximately equal to the whole lung volume, it corresponds to the average number of open terminal airways. If we consider that n generations are present, we can easily show that $V \propto P^n$. This result is found to be in very good agrement with experiments (Suki *et al.*, 1998).

11
Avalanches in Biological Systems

In the present book, we have discussed several examples of avalanche phenomena and crackling noise derived from the physics of materials. Similar phenomena, however, are also observed in living systems. We provide here a few representative examples to illustrate this point. In particular, we consider the burst dynamics of punctuated evolution, the abrupt phenotype transformations controlled by gene regulatory networks in the cell, and the intermittent motion of collective cell migration which is relevant for cancer invasion. This list of examples is far from exhaustive but it does provide an idea of the relevance of avalanche phenomena for biological systems.

11.1 Punctuated evolution

According to the Darwin's theory of evolution, species slowly evolve by random mutations and by selection of the fittest. Observing fossil records, Gould and Eldredge suggested that evolution does not proceed gradually but is instead characterized by long periods of stasis without changes in species followed by short and rapid evolution events (Gould and Eldredge, 1977). In other words, this theory of *punctuated equilibria* states that evolution occurs in avalanches. A quantitative analysis of the extinction record confirms that the number of extinction events of a given size decays as a power law with an exponent $\tau \simeq 2$ (see Fig. 11.1) (Sneppen *et al.*, 1995; Newman, 1996; Solé and Manrubia, 1996; Sole *et al.*, 1997), suggesting that evolution may be described by a (self-organized) critical process (Bak and Sneppen, 1993; Sneppen *et al.*, 1995).

Bak and Sneppen (1993) proposed a toy model to illustrate how a power law distribution of events might arise in an evolutionary context. The Bak–Sneppen (BS) model considers a set of N species characterized by a fitness value f_i, with $i = 1, \ldots N$. Fitness values are initially drawn from a uniform distribution in the interval $[0, 1]$. The evolutionary process is simulated by eliminating the species with the smallest fitness and replacing it with a new species with a new fitness extracted from the same uniform distribution. The authors imagine that the extinction of the species i may provoke the extinction of other species that depend on it. For instance, it was imagined that species $i + 1$ and $i - 1$ are dependent on species i, so if i becomes extinct and it is replaced then species $i + 1$ and $i - 1$ also become extinct and their fitness is replaced by a new random number.

The extremal dynamics of the BS model is reminiscent of invasion percolation (see section 3.2) where sites with the minimum threshold are selected for invasion. Avalanches can be defined as in invasion percolation from the time-ordered sequence of the species that become extinct (see Fig. 3.5 and replace the threshold x_i with the

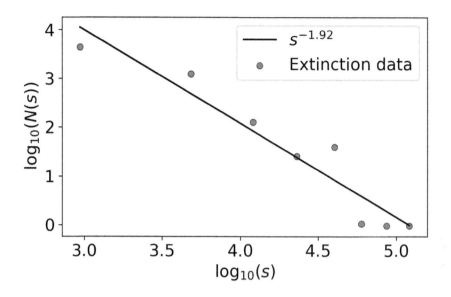

Fig. 11.1 The number of extinction events of a given size. [Data from (Newman, 1996)].

fitness value f_i). As the dynamics proceeds, sites with low fitness values are selected for extinction so that the distribution of f_i develops a gap that converges to an asymptotic distribution $[f_c, 1]$ (Paczuski *et al.*, 1996). The value f_c corresponds to a critical point and avalanches are then power-law distributed with an exponent $\tau \simeq 1.1$. It is also possible to define a mean-field version of the model where instead of selecting nearest-neighbor sites after an extinction one selects k other random sites (Flyvbjerg *et al.*, 1993). In this case, as for other mean-field models the avalanche distribution decays as a power law with exponent $\tau = 3/2$ (Flyvbjerg *et al.*, 1993).

The BS model illustrates how a simplified model of evolution based on species selection and interactions may lead to a power-law distribution of events that is in qualitative agreement with fossil records. This is, however, not the only explanation for the same data, and other mechanisms and models have been proposed in the literature (Newman, 1996; Solé and Manrubia, 1996). The idea that evolution might follow a critical dynamics is still debated (Sole and Bascompte, 1996; Kirchner and Weil, 1998) since one can easily devise models that generate power-law-distributed events which do not involve critical points (Newman, 1996).

11.2 Gene regulatory networks

The genetic code inscribed in DNA contains all the information necessary to build proteins, which are essential components of cells, responsible for most of their activity. While a cell has a unique genetic code, its actual state, or phenotype, can change in time depending for instance on how many proteins of each kind are present at a given instant. In fact, not all of the genes are active at the same time and in the same way. The way cells activates genes and produces proteins at any given time in response to external and internal stimuli is ruled by a complex regulatory network. Genes are transcribed into messenger RNA by a complex machinery employing the enzyme RNA-polymerase. Messenger RNA translocates to the ribosomes that translate their information content into a sequence of amino acids that then fold into proteins.

How much messenger RNA is actually transcribed at each time depends on transcription factors, special proteins devoted to this function, but also on a host of other mechanisms too complicated to describe here. At a very simplistic level, we can imagine that each gene sits at a node of the network and is connected to some other nodes by directed edges indicating the regulatory interactions that exist among the genes. For instance, gene A encodes a transcription factor that might induce the expression of gene B, which in its turn might suppress the activity of gene C and enhance that of gene D. Considering all the possible interactions between genes, it is possible to draw a complex gene regulatory network that would describe how the state of a cell changes in time. Fig. 11.2 displays for instance a reconstruction of the gene regulatory network of a yeast cell (Ma *et al.*, 2014).

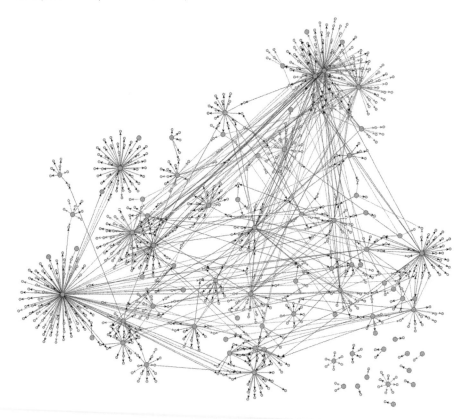

Fig. 11.2 Gene regulatory network of yeast *Saccharomyces cerevisiae*. Each node represents a gene and each arrow a regulatory interaction. Black arrows are activating and gray arrows are repressing. Source: From (Ma *et al.*, 2014), CC 4.0 license.

The expression level of each gene in the network can be quantified as the number of RNA molecules corresponding to each gene that are present in the cell at a given time, which is also known as the transcriptome. We can consider the transcriptome as a proxy for the cell phenotype. According to an old and influential metaphor due to Waddington (1957), we may view the cell phenotype as a marble rolling over an *epigenetic landscape*. In this scenario, phenotypic transformations correspond to the marble crossing a hill separating different valleys. The Waddington landscape should correspond to the attractors or fixed points of the kinetics of the gene regulatory network of the cell (Kauffman 1969; Li *et al.* 2004; Huang *et al.* 2005, 2007; Wang *et al.* 2010, 2011; Huang 2012; Li and Wang 2013). One could then try to map experimentally measured gene expression data to the attractor states of the network (Scialdone *et al.* 2016; Bargaje *et al.* 2017; Font-Clos *et al.* 2018, 2021).

The level of expression of the genes change all the time in response to external and internal perturbations, such as external stimuli due to the environment or random mutations. It is possible to perform experiments to artificially change the expression level of a gene. When the expression level is suppressed, one talks about *gene knockdown* and when it is increased one talks about *gene overexpression*. Biologists routinely perform these kinds of experiments to investigate the role of each of the many genes—around 20,000 in humans—present in the DNA. Since genes do not work in isolation, knockdown or overexpression of an individual gene may perturb the expression level of the genes it is connected to within the gene regulatory network. This may lead to a cascade event or an avalanche of gene-expression changes. Researchers have tested this idea systematically in a relatively simple organism like yeast by knocking down each one of its genes and then measuring the associated gene expression changes (Hughes *et al.*, 2000). From these data, it is possible to define the size of an avalanche by counting how many genes changed their expression above a threshold in response to a single knockdown (Serra *et al.*, 2004). The corresponding avalanche distribution reported in Fig. 11.3 displays a power-law behavior with exponent $\tau \simeq 1.1$. The statistical sampling is too small to reach a firm conclusion on the presence of a cutoff other than the system size (i.e. the total number of genes in yeast which is more than 6000).

It is in principle possible to model the dynamics of gene regulatory networks by writing coupled kinetic reaction equations for all the nodes in the network. The main issue is that estimating experimentally the value of the reaction rates is a daunting task that has been performed only in some limited cases (see e.g. Lee *et al.* 2003). A way out is to resort to simplified Boolean network models where the expression levels of the genes in the network can only assume a binary value $s_i = 0, 1$ (Kauffman, 1969; Kauffman, 1993). In random Boolean networks, each one of the N nodes is connected to K randomly selected input nodes and the Boolean value of each node is selected according to randomly chosen Boolean functions (Kauffman, 1993). The number of output nodes A is a not constant but its average must obey $\langle A \rangle = K$. Using this model to simulate knockdown experiments, it is possible to obtain avalanches whose size distribution is in reasonable agreement with experiments (Serra *et al.*, 2007). Detailed studies of random Boolean networks show that the model attractors display a transition between a stable phase and a chaotic phase controlled by the parameter K. Critical behavior is expected for $K = K_c$. Observation of power-law scaling for

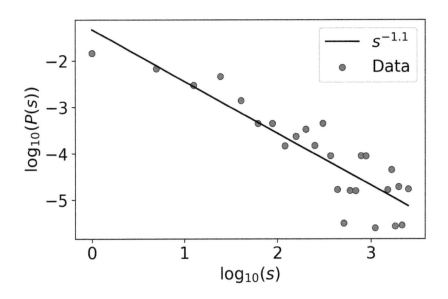

Fig. 11.3 The distribution of avalanche sizes associated with the knockdown of individual genes in yeast. The line is a simple power-law fit of the data. Source: Experimental data from Hughes *et al.* (2000), extracted from Serra *et al.* (2004) Fig. 1.

knock-down experiments suggests that gene regulatory networks are somehow close to a critical point. There is still not a commonly accepted explanation for this critical behavior. One point to keep in mind is that real regulatory networks are not random but *minimally frustrated* (Tripathi *et al.*, 2020). In the language of Boolean networks, a frustrated circuit is one where it is not possible to satisfy all the Boolean functions at the same time. Random networks are not guaranteed to be devoid of frustration.

It is important to remark that biological processes in the cell do not depend on the entire regulatory network, but often only on a small subset of genes that are said to be part of a *pathway*. Each pathway can also be described by a network and biologists have created large databases with pathways relevant for the most important cellular processes. For example, epithelial (E) cells can change into mesenchymal (M) cells through an epithelial-to-mesenchymal transition (EMT), associated with the loss of cell–cell adhesion and the gain of invasive traits. Since the EMT is particularly relevant for metastatic tumors, a great effort has been devoted to identifying the pathways that regulate it. Evidence shows that the EMT is a gradual process where the cell can acquire hybrid E/M phenotypes (Bitterman *et al.*, 1990; Haraguchi *et al.*, 1999; Paniz Mondolfi *et al.*, 2013), which in tumors are associated with a poor patient outcome (Jolly *et al.*, 2016; George *et al.*, 2017; Font-Clos *et al.*, 2021).

Pathways regulating the EMT have been simulated considering minimal switches composed of few genes (Jolly *et al.*, 2014) and large networks with Boolean rules (Steinway *et al.* 2014, 2015; Font-Clos *et al.* 2018, 2021) and continuous dynamics (Huang *et al.*, 2017). We illustrate here one example where a pathway composed of more than 70 genes regulating the EMT was simulated with the following evolution rule (Font-Clos *et al.*, 2018)

$$s_i(t + 1) = \text{sign}\left(\sum_j J_{ij}s_j(t)\right), \tag{11.1}$$

where $J_{ij} = \pm 1$ depending on the type of interaction that node j has with node i. For activating interactions we have $J_{ij} = 1$, while for inhibiting interactions we ha have $J_{ij} = -1$. Notice the formal similarity between Eq. 11.1 and Eq. 3.7 introduced in Chapter 3 to simulate the zero-temperature dynamics in the RFIM (Sethna *et al.*, 1993) and in spin glasses (Pázmándi *et al.*, 1999). In analogy with magnetic spin systems, we could define a pseudo-Hamiltonian $\mathcal{H} = -\sum_{i,j} J_{ij} s_i s_j$, although here $(J_{ij} \neq J_{ji})$ It is only when interactions are strictly symmetric $(J_{ij} = J_{ji})$, as in magnetic systems, that the fixed points of Eq. 11.1 are local minima of $\mathcal{H}$ (Derrida, 1987; Gutfreund *et al.*, 1988). In the case of gene regulatory networks, this is not guaranteed, but simulations show that $\mathcal{H}$ is lowered under repeated application of Eq. 11.1, thus providing a measurement of the stability of the states of the network (Font-Clos *et al.*, 2018).

Fig. 11.4 shows a picture of the pseudo-energy landscape obtained from a large number of simulations starting from random initial conditions and evolved until the network state reaches a fixed point. The two valleys represent E and M states and the peak hybrid E/M states. One can then investigate how the landscape changes when each one of the nodes is held fixed to $s_i = \pm 1$, which simulates overexpression or knock-down of the corresponding gene (Font-Clos *et al.*, 2018). As an example, Fig 11.4(b) reports the trajectories of the system initially placed into an E state after the over-expression of the gene SNAIL1, which is known to be relevant for the EMT. The trajectory crucially depends on the initial state, and states with high $\mathcal{H}$ are most likely to undergo EMT. This is confirmed by measuring the distribution of the number of nodes s affected by the knock-down or overexpression of a single gene (see Fig. 11.4(c)). The distribution decays as power law $P(s) \sim s^{-\tau}$ up to a cutoff value that increases with the value of $\mathcal{H}$ in the initial state (see Fig. 11.4(d)), a further indication that high-$\mathcal{H}$ states are more susceptible to fluctuations. The avalanche exponent of the power law distribution is $\tau \simeq 3/2$, suggesting that the model is described by mean-field theory (Font-Clos *et al.*, 2018).

11.3 Collective cell migration

Cells in tissues often move collectively (Friedl and Gilmour, 2009), a process of important biological relevance for cancer metastasis (La Porta and Zapperi, 2017), where clusters of cells can invade neighboring tissues together (Rørth, 2009; Khalil and Friedl, 2010; Gov, 2014), and in wound healing where cells should move together to restore a damaged tissue. A detailed review about cell migration can be found in La Porta and Zapperi (2019). We can consider cell assemblies as a non-equilibrium active form of matter, owing their dynamic properties to the conversion of biochemical energy into kinetic energy (Ramaswamy, 2010). In analogy with ordinary materials, active matter displays phase transitions among different phases characterized by distinct structural

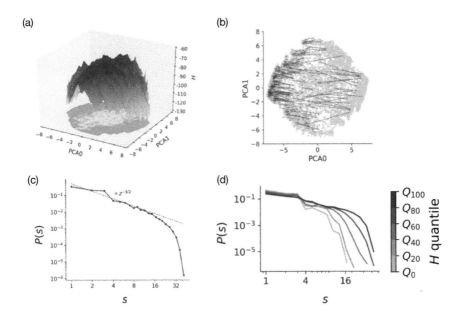

Fig. 11.4 Avalanches in the EMT. (a) The landscape associated with the gene regulatory network associated with the EMT. (b) Possible transitions associated with the knock-down of a single gene. (c) Distribution of avalanche sizes. (d) The cutoff of the distribution of avalanches size depends on the value of the pseudo-energy of the initial state. Source: (Font–Clos *et al.*, 2018).

and dynamic properties (La Porta and Zapperi, 2020). While at low concentration cells are able to move freely in a persistent random motion, at high concentrations mutual cell interactions slow down their motion. At very high concentrations, cells jam and their motion is confined as with the molecules of an ordinary glass (Angelini *et al.*, 2010; Park *et al.*, 2015; Malinverno *et al.*, 2017). There is, however, a tug-of-war between the cells' ability to move and the the caging effects due to crowding, and if cell propulsion prevails cells enter into a collective flow state. This is reminiscent of the jamming/unjamming transition, widely observed in soft-matter systems such as foams, colloids, or granular media (Liu *et al.*, 2010) and discussed in section 6.5 in the context of plastic yielding.

Contrary to ordinary matter, unjamming in cellular assemblies is driven by internal self-propulsion forces which are activated when cells are faced with an empty space, a condition that can be reproduced *in vitro* by wound-healing assays (Vedula *et al.*, 2013; Szabó *et al.*, 2006; Poujade *et al.*, 2007; Sepúlveda *et al.*, 2013). In these experiments, a confined monolayer of cells is scratched, creating an empty space whose invasion can be observed in time-lapse microscopy. In these conditions, the dynamics is controlled by a complex network of biochemical pathways (Ilina and Friedl, 2009) but also the physical interactions among cells and with their environment are also known to play a role. For instance, the cell migration velocity depends on the composition and stiffness of the extracellular matrix (Tambe *et al.*, 2011; Brugues *et al.*, 2014; Haeger *et al.*, 2014; Lange and Fabry, 2013; Koch *et al.*, 2012). Furthermore, cells are able to transfer mechanical stresses to their neighbors (Tambe *et al.*, 2011), producing long-ranged stress waves in the monolayer (Serra-Picamal *et al.*, 2012; Banerjee *et al.*, 2015). This observation suggests an analogy with the disordered elastic systems treated in Chapter 4, where the dynamics is ruled by the interplay of elastic interactions and the interaction with a quenched random force field. In the case of collective cell migration, the elastic interactions are provided by intracellular adhesion, while random forces are due to the substrate.

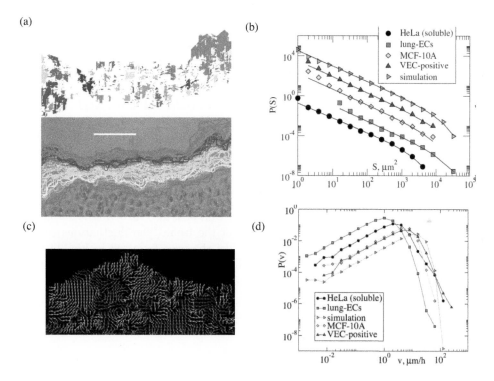

Fig. 11.5 Dynamic fluctuations in wound healing experiments. (a) An example of cell a cell front and the activity maps reconstructed from the time evolution of the front in HeLa cells moving on a collagen substrate. Regions marked by the same color in the activity map move collectively. The scale bar is 100 μm. (b) The distributions of the areas of activity clusters can be described by a power law with a cutoff. The distributions for different cell types have been shifted for clarity. The slope obtained fitting the distributions is very similar for all cell types. (c) Velocity map obtained from particle image velocimetry. The length of the arrows is proportional to the magnitude of the velocity. (d) Distributions of velocity magnitudes for different cell types. Source: Reprinted from Chepizhko *et al.* (2016) with permission.

Indeed, a careful analysis of time-lapse imaging during wound healing confirms that a migrating cell front shares many similarities with a moving front close to the depinning transition (Chepizhko *et al.* 2016, 2018). The analysis has been performed on a variety of cell lines (human cancer cells and epithelial cells, mouse endothelial cells) over different substrates (plastic, soluble, and fibrillar collagen) and with varying experimental conditions (e.g. the tuning of intracellular adhesion by VE-cadherin knock down). An example of the evolution of the cell front in a monolayer of HeLa cells is reported in the bottom part of Fig. 11.5(a). The fronts are rough and advance in bursts, as is apparent by looking at the activity map in the top part of Fig. 11.5(a), where the shaded region corresponds to areas that move collectively, denoted as clusters of activity. Activity maps were obtained using the same method discussed in section 5.4 in the context planar crack propagation (Tallakstad *et al.*, 2011) and later applied also to imbibition (Clotet *et al.*, 2014). As in the case of fracture or imbibition, the distribution of cluster areas S decays as a power law $P(S) \sim S^{-\tau}$ up to a cutoff length S^*, as illustrated in 11.5(b) for a variety of cell lines. It is interesting to remark that the value of the exponent $\tau \simeq 1.5$ is independent of the cell line (Chepizhko *et al.*, 2016) and is similar to the one observed in fracture (Tallakstad *et al.*, 2011). It was also shown that all the avalanche size distributions can be collapsed into a single universal scaling function (Chepizhko *et al.*, 2016).

In addition to the activity map, a useful technique to characterize the fluctuations in the dynamics of cell migration, both in confluent and in wound healing conditions, is provided by particle image velocimetry (PIV). PIV estimates local velocities by performing a digital image correlation analysis on the time-lapse sequence and allows to obtain a velocity map, such as the one shown in Fig. 11.5(c) for HeLa cells. The figure shows that cells move with significant fluctuations, also involving local motion that is opposite to the propagation direction of the front. The fluctuations can be captured by measuring velocity distributions like the ones reported in Fig. 11.5(d). The distribution varies slightly for different cell lines, but the shape of the distribution is similar in all cases, again suggesting universal scaling.

12
Outlook

In this book, we have presented an overview of crackling noise and have discussed its interpretation in terms of underlying avalanche phenomena. We have seen that the statistical properties of crackling noise are characterized by a set of power-law distributions which can be related by scaling laws in analogy with similar relations derived for equilibrium- and non-equilibrium-critical systems. This suggests that crackling noise is a manifestation of some for of criticality of the avalanches that produce the noise. While in this book we followed the general idea that power-law distributions in crackling noise are due to critical avalanches, several other mechanisms not involving critical points have been proposed in the literature. However, none of these mechanisms of power-law generation have gained much traction in the context of crackling noise. While not all power laws in nature are a signature of a critical point, avalanche phenomena have been consistently and quantitatively interpreted using critical phenomena.

Throughout the book we have illustrated different types of critical points that can give rise to avalanches and crackling noise. The main issue with this interpretation is that critical points in principle need fine-tuning so that generically one would not expect observe scaling except very close to the critical point. It is important to remember that power laws in nature are always observed over a finite number of scales, up to a cut-off. Hence, the central question becomes: What determines the cut-off scale in crackling noise? Is it just due to the lack of statistical sampling? In that case, the power law would continue to larger scales if we just had more data. Is it due to a finite size effect? This would mean that if we had a larger system, we could extend the scaling regime. Otherwise, the cutoff could reflect more fundamental physical aspects of the problem which would make the system only near-critical.

We recall shortly the main mechanisms that could be relevant in order to explain the appearance of a cut-off in the avalanche distributions. In some driven disordered systems exemplified by the RFIM, the critical region is rather large, so one might observe extended scaling even relatively far away from the critical point (Perkovic et al., 1995) (see Chapter 3). Another alternative explanation lies in self-organized criticality (Bak et al., 1987) which implies that the system is driven by its dynamics to the close vicinity of an otherwise fine-tuned critical point. This can happen when the control parameter of the phase transition is dynamically coupled to the order parameter by a feedback loop, as proposed early on in Sornette (1992a). We have illustrated this point in the context of interface depinning where the driving force F could be coupled to the interface position u as $F = k(Vt - u)$. In the limit $V \to 0$ and $k \to 0$, the avalanches are power-law distributed over all length scales (see Chapter 4). This mechanism takes place in the Barkhausen effect thanks to the coupling provided by the demagnetizing field (see Chapter 8), in plasticity where the coupling is due to strain hardening (see Chapter 6), and in granular friction thanks to the stiffness of apparatus (see Chapter 7). Finally, power laws might occur because the system is driven towards the critical point by a slow increase of the control parameter, as first discussed in Sornette 1994. If no feedback mechanism is in place, then the crackling noise would turn out to be non-stationary with avalanches starting small and then becoming larger as the critical point is approached. We have seen an example of this behavior when we discussed AE in fracture (see Chapter 5). Depending on the type of critical behavior, we can expect different scaling relations to apply. For instance, when the system is driven towards the critical point in a non-stationary way, the avalanche exponent is larger than in the case of the stationary critical point (see for instance the discussions in sections 3.4 and 4.5.2).

The fact that crackling noise statistics could be interpreted in terms of some type of critical phenomenon has far-reaching implications. Critical phenomena are expected to be universal so that the value of the critical exponent does not depend on microscopic details but only on symmetries and conservation laws. We have seen a practical example of this when discussing the Barkhausen noise, where universal scaling was found to describe broad classes of ferromagnetic alloys with amorphous or polycrystalline structures (see Chapter 8). Universality also implies that not only exponents, but also entire scaling functions, are the same for structurally different materials. We have provided several examples of this fact, from the average pulse shape in crackling noise to the cutoff function in avalanche distributions. We could say that it is universality that allows theory to be successful. The fact that some physical observable does not depend on all the microscopic details of the system allows us to devise simplified models which provide quantitatively correct predictions.

In this book, we have provided several examples of crackling noise, mostly drawn from the physics of disordered materials. While these examples illustrate the main mechanisms underlying avalanche phenomena, the list is far from the exhaustive, and many more examples exist in widely different phenomena. For instance, in geophysical phenomena one can find large-scale application of some basic physical processes that we have discussed here such as fracture, friction and plasticity. The most dramatic example is provided by earthquakes, whose power-law amplitude distribution is widely known as the *Gutemberg–Richter* law (Gutenberg, 1954). Whether the statistics of earthquakes can be explained with any of the mechanisms discussed in the present book is still open for debate, with several articles and books devoted to the subject (see for instance Hergarten 2002). Interesting examples of avalanches in the geophysical setting also include landslides (Chen *et al.*, 2007), rockfalls (Dussauge *et al.*, 2003), and snow avalanches (Birkeland and Landry, 2002; Faillettaz *et al.*, 2004). Other large-scale avalanche observations include solar flares (Aschwanden and Güdel, 2021) and magnetospheric plasmas (Klimas *et al.*, 2000).

Finally, we have touched upon a few examples of crackling noise in biological systems, such as the dynamics of punctuated evolution, the fluctuations in gene regulatory networks, and the front propagation in collective cell migration (see Chapter 11). Several other interesting examples exist as recently reviewed in Munoz (2018). We mention here for instance the observation of power-law-distributed avalanches in cytoskeletal rearrangements (Floyd *et al.*, 2021) or in neuronal activity (Beggs and Plenz, 2003; Petermann *et al.*, 2009; Klaus *et al.*, 2011). The observation of neuronal avalanches has been extremely influential in neurosciences, leading to speculation that the brain might operate close to a critical point (Plenz *et al.*, 2021). Interestingly, self-organized criticality in the brain has been explained invoking synaptic plasticity (Zeraati *et al.*, 2021), which provides a dynamic feedback loop similar to the one extensively discussed in this book. All these examples show that crackling noise and avalanche phenomena are very common in living systems; the tools and concepts discussed in this book could therefore be useful to understand a large variety of phenomena in a wide diversity of areas.

Bibliography

Abrikosov, A. A. (2004). Nobel lecture: Type-II superconductors and the vortex lattice. *Rev. Mod. Phys.*, **76**(3), 975.

Aegerter, C. M. (1998). Evidence for self-organized criticality in the bean critical state in superconductors. *Phys. Rev. E*, **58**(2), 1438.

Aegerter, C. M., Gunther, R., and Wijngaarden, R. J. (2003). Avalanche dynamics, surface roughening, and self-organized criticality: Experiments on a three-dimensional pile of rice. *Phys. Rev. E*, **67**(5), 051306.

Aegerter, C. M., Lorincz, K. A., Welling, M. S., and Wijngaarden, R. J. (2004). Extremal dynamics and the approach to the critical state: Experiments on a three-dimensional pile of rice. *Phys. Rev. Lett.*, **92**(5), 058702.

Alava, M., Rost, M., and Dubé, M. (2004). Imbibition in disordered media. *Advances in Physics*, **53**, 83–175.

Alava, M. J. (2002). Scaling in self-organized criticality from interface depinning? *J. Phys.: Condensed Matter*, **14**(9), 2353.

Alava, M. J. and Lauritsen, K. B. (2001). Quenched noise and over-active sites in sandpile dynamics. *Europhys. Lett.*, **53**(5), 563.

Alava, M. J. and Munoz, M. A. (2002). Interface depinning versus absorbing-state phase transitions. *Phys. Rev. E*, **65**(2), 026145.

Alava, M. J., Nukala, P. K. V. V., and Zapperi, S. (2006). Statistical models of fracture. *Advances in Physics*, **55**(3-4), 349–476.

Alencar, A. M., Hantos, Z., Petak, F., Tolnai, J., Asztalos, T., Zapperi, S., Andrade Jr., J. S., Buldyrev, S. V., Stanley, H. E., and Suki, B. (1999). Scaling behavior in crackle sound during lung inflation. *Phys. Rev. E*, **60**(4), 4659–63.

Alessandro, B., Beatrice, C., Bertotti, G., and Montorsi, A. (1990*a*). Domain wall dynamics and Barkhausen effect in metallic ferromagnetic materials. i. Theory. *J. Appl. Phys.*, **68**, 2901–08.

Alessandro, B., Beatrice, C., Bertotti, G., and Montorsi, A. (1990*b*). Domain wall dynamics and Barkhausen effect in metallic ferromagnetic materials. ii. Experiments. *J. Appl. Phys.*, **68**, 2908–15.

Ali, A. A. and Dhar, D. (1995). Breakdown of simple scaling in Abelian sandpile models in one dimension. *Phys. Rev. E*, **51**(4), R2705.

Alstrøm, P. (1988). Mean-field exponents for self-organized critical phenomena. *Phys. Rev. A*, **38**(9), 4905.

Altshuler, E. and Johansen, T. H. (2004). Experiments in vortex avalanches. *Rev. Mod. Phys.*, **76**, 471.

Altshuler, E., Johansen, T. H., Paltiel, Y., Jin, P., Bassler, K. E., Ramos, O., Chen, Q., Reiter, G. F., Zeldov, E., and Chu, C. (2004). Vortex avalanches with robust statistics observed in superconducting niobium. *Phys. Rev. B*, **70**(14), 140505.

Amaral, L. A. N., Barabási, A.-L., Buldyrev, S. V., Harrington, S. T., Havlin, S., Sadr-Lahijany, R., and Stanley, H. E. (1995). Avalanches and the directed percolation depinning model: Experiments, simulations, and theory. *Phys. Rev. E*, **51**(5), 4655–73.

Amaral, L. A. N. and Lauritsen, K. B. (1996). Self-organized criticality in a rice-pile model. *Phys. Rev. E*, **54**(5), R4512.

Ananthakrishna, G. (2007). Current theoretical approaches to collective behavior of dislocations. *Physics Reports*, **440**, 113–259.

Ananthakrishna, G., Noronha, S. J., Fressengeas, C., and Kubin, L. P. (1999). Crossover from chaotic to self-organized critical dynamics in jerky flow of single crystals. *Phys. Rev. E*, **60**, 005455.

Andersen, J. V., Sornette, D., and Leung, K. W. (1997). Tricritical behavior in rupture induced by disorder. *Phys. Rev. Lett.*, **78**, 2140.

Angelini, T. E., Hannezo, E., Trepat, X., Fredberg, J. J., and Weitz, D. A. (2010). Cell migration driven by cooperative substrate deformation patterns. *Phys. Rev. Lett.*, **104**(16), 168104.

Antonaglia, J., Wright, W. J., Gu, X., Byer, R. R., Hufnagel, T. C., LeBlanc, M., Uhl, J. T., and Dahmen, K. A. (2014*a*). Bulk metallic glasses deform via slip avalanches. *Phys. Rev. Lett.*, **112**(15), 155501.

Antonaglia, J., Xie, X., Schwarz, G., Wraith, M., Qiao, J., Zhang, Y., Liaw, P. K., Uhl, J. T., and Dahmen, K. A. (2014*b*). Tuned critical avalanche scaling in bulk metallic glasses. *Scientific reports*, **4**(1), 1–5.

Aranson, I. S. and Tsimring, L. S. (2006). Patterns and collective behavior in granular media: Theoretical concepts. *Rev. Mod. Phys.*, **78**(2), 641.

Arbabi, S. and Sahimi, M. (1990). Test of universality for three-dimensional models of mechanical breakdown in disordered solids. *Phys. Rev. B*, **41**, 772.

Arbabi, S. and Sahimi, M. (1993). Mechanics of disordered solids. i. percolation on elastic networks with central forces. *Phys. Rev. B*, **47**(**2**), 695–702.

Argon, A. and Kuo, H. (1979). Plastic flow in a disordered bubble raft (an analog of a metallic glass). *Materials Science and Engineering*, **39**(1), 101 – 109.

Argon, A. S. (1979). Plastic deformation in metallic glasses. *Acta metallurgica*, **27**(1), 47–58.

Arndt, P. F. and Nattermann, T. (2001). Criterion for crack formation in disordered materials. *Phys. Rev. B*, **63**, 134204.

Aschwanden, M. J. and Güdel, M. (2021). Self-organized criticality in stellar flares. *The Astrophysical Journal*, **910**(1), 41.

Aström, J. A., Alava, M. J., and Timonen, J. (2000). Roughening of a propagating planar crack front. *Phys. Rev. E*, **62**, 2878.

Bagnold, R. A. (1966). The shearing and dilatation of dry sand and the "singing" mechanism. *Proc. R. Soc. London A*, **295**, 219–32.

Bahiana, M., Koiller, B., de Queiroz, S. L. A., Denardin, J. C., and Sommer, R. L. (1999). Domain size effect in Barkhausen noise. *Phys. Rev. E*, **59**, 3884–7.

Bailey, N. P., Schiøtz, J., Lemaître, A., and Jacobsen, K. W. (2007). Avalanche size scaling in sheared three-dimensional amorphous solid. *Phys. Rev. Lett.*, **98**(9), 095501.

Bak, P. and Sneppen, K. (1993). Punctuated equilibrium and criticality in a simple model of evolution. *Phys. Rev. Lett.*, **71**(24), 4083–6.

Bak, P., Tang, C., and Wiesenfeld, K. (1987). Self-organized criticality: An explanation of the 1/f noise. *Phys. Rev. Lett.*, **59**, 381–4.

Bakó, B., Groma, I., Györgyi, G., and Zimányi, G. T. (2007). Dislocation glasses: Aging during relaxation and coarsening. *Phys. Rev. Lett.*, **98**(7), 075701.

Baldassarri, A., Dalton, F., Petri, A., Zapperi, S., Pontuale, G., and Pietronero, L. (2006). Brownian forces in sheared granular matter. *Phys. Rev. Lett.*, **96**(11), 118002.

Balents, L., Marchetti, M. C., and Radzihovsky, L. (1997). Comment on ɪmoving glass phase of driven latticesɟ. *Phys. Rev. Lett.*, **78**(4), 751.

Banerjee, S., Utuje, K. J. C., and Marchetti, M. C. (2015). Propagating stress waves during epithelial expansion. *Phys. Rev. Lett.*, **114**(22), 228101.

Barabasi, A.-L., Buldyrev, S. V., Stanley, H. E., and Suki, B. (1996). Avalanches in the lung: A statistical mechanical model. *Phys. Rev. Lett.*, **76**(12), 2192–2195.

Baret, J.-C., Vandembroucq, D., and Roux, S. (2002). Extremal model for amorphous media plasticity. *Phys. Rev. Lett.*, **89**(19), 195506.

Barford, W. (1997). Avalanches in the bean critical-state model. *Phys. Rev. B*, **56**, 435.

Barford, W., Beere, W. H., and Steer, M. (1993). The dynamics of the bean critical state. *J. Phys.: Condens. Matter*, **5**, L333.

Bargaje, R., Trachana, K., Shelton, M. N., McGinnis, C. S., Zhou, J. X., Chadick, C., Cook, S., Cavanaugh, C., Huang, S., and Hood, L. (2017). Cell population structure prior to bifurcation predicts efficiency of directed differentiation in human induced pluripotent cells. *PNAS*, **114**(9), 2271–2276.

Barthelemy, M., da Silveira, R., and Orland, H. (2002). The random fuse network as a dipolar magnet. *Europhys. Lett.*, **57**, 831–837.

Bassler, K. E. and Paczuski, M. (1998). Simple model of superconducting vortex avalanches. *Phys. Rev. Lett.*, **81**, 3761.

Batrouni, G. G., Kahng, B., and Redner, S. (1988). Conductance and resistance jumps in finite-size random resistor networks. *J. Phys. A: Mathematical and General*, **21**(1), L23–L29.

Bean, C. P. (1962). Magnetization of hard superconductors. *Phys. Rev. Lett.*, **8**, 250.

Beggs, J. M. and Plenz, D. (2003). Neuronal avalanches in neocortical circuits. *Journal of neuroscience*, **23**(35), 11167–11177.

Behnia, K., Capan, C., Mailly, D., and Etienne, B. (2000). Internal avalanches in a pile of superconducting vortices. *Phys. Rev. B*, **61**, 3815.

Benassi, A. and Zapperi, S. (2011). Barkhausen instabilities from labyrinthine magnetic domains. *Phys. Rev. B*, **84**(21), 214441.

Berger, A., Inomata, A., Jiang, J. S., Pearson, J. E., and Bader, S. D. (2000). Experimental observation of disorder-driven hysteresis-loop criticality. *Phys. Rev. Lett.*, **85**, 4176.

Berkowitz, B. and Ewing, R. P. (1998). Percolation theory and network modeling applications in soil physics. *Surveys in Geophysics*, **19**(1), 23–72.

Bertotti, G. (1986). Statistical interpretation of magnetization processes and eddy current losses in ferromagnetic materials. In *Proc. of the 3rd Int. Conf. Physics of Magnetic Materials* (ed. W. Gorzkowski, H. Lachowicz, and H. Szymczak), Singapore, pp. 489–508. Word Scientific.

Bertotti, G. (1987). Dynamics of magnetic domain walls and Barkhausen noise in metallic ferromagnetic systems. In *Magnetic Excitations and Fluctuations II* (ed. U. Balucani, S. W. Lovesey, M. G. Rasetti, and V. Tognetti), pp. 135–139. Springer,Berlin.

Bertotti, G. and Pasquale, M. (1990). Hysteresis phenomena and Barkhausen-like instabilities in the Sherrington-Kirkpatrick spin glass model. *J. Appl. Phys.*, **67**, 5255–5257.

Beyerlein, I. J. and Phoenix, S. L. (1996). Stress concentrations around multiple fiber breaks in an elastic matrix with local yielding or debonding using quadratic influence superposition. *J. Mech. Phys. Solids*, **44**, 1997.

Birkeland, K. and Landry, C. (2002). Power-laws and snow avalanches. *Geophysical Research Letters*, **29**(11), 49–1.

Bishop, J. E. L. (1980). The contribution made by eddy currents to the effective mass of a magnetic domain wall. *J. Phys. D: Appl. Phys.*, **13**, L15–L19.

Bitterman, P., Chun, B., and Kurman, R. J. (1990). The significance of epithelial differentiation in mixed mesodermal tumors of the uterus. a clinicopathologic and immunohistochemical study. *Am J Surg Pathol*, **14**(4), 317–28.

Blatter, G., Feigelman, M. V., Geshkenbein, V. B., Larkin, A. I., and Vinokur, V. M. (1994). Vortices in high-temperature superconductors. *Rev. Mod .Phys.*, **66**, 1125.

Blumberg, S., Frank, O., and Hofmann, T. (2010). Quantitative studies on the influence of the bean roasting parameters and hot water percolation on the concentrations of bitter compounds in coffee brew. *Journal of agricultural and food chemistry*, **58**(6), 3720–3728.

Bonachela, J. A., Alava, M., and Muñoz, M. A. (2009). Cusps, self-organization, and absorbing states. *Phys. Rev. E*, **79**(5), 050106.

Bonamy, D., Ponson, L., Prades, S., Bouchaud, E., and Guillot, C. (2006). Scaling exponents for fracture surfaces in homogeneous glass and glassy ceramics. *Phys. Rev. Lett.*, **97**(13), 135504.

Bonamy, D., Santucci, S., and Ponson, L. (2008). Crackling dynamics in material failure as the signature of a self-organized dynamic phase transition. *Phys. Rev. Lett.*, **101**(4), 045501.

Bonfanti, S., Ferrero, E. E., Sellerio, A. L., Guerra, R., and Zapperi, S. (2018). Damage accumulation in silica glass nanofibers. *Nano letters*, **18**(7), 4100–4106.

Bonfanti, S., Guerra, R., Mondal, C., Procaccia, I., and Zapperi, S. (2019). Elementary plastic events in amorphous silica. *Phys. Rev. E*, **100**(6), 060602.

Bouchaud, J. P., Bouchaud, E., Lapasset, G., and Planès, J. (1993). Models of fractal cracks. *Phys. Rev. Lett.*, **71**, 2240.

Bouchbinder, E., Mathiesen, J., and Procaccia, I. (2004). Roughening of fracture surfaces: The role of plastic deformation. *Phys. Rev. Lett.*, **92**, 245505.

Brailsford, A. (1965). Effective line tension of a dislocation. *Phys. Rev.*, **139**(6A), A1813.

Brandt, E. H. (1995). The flux-line lattice in superconductors. *Rep. Prog. Phys.*, **58**,

1465.

Bray, A. J. (2002). Theory of phase-ordering kinetics. *Advances in Physics*, **51**(2), 481–587.

Bretz, M., Cunningham, J. B., Kurczynski, P. L., and Nori, F. (1992). Imaging of avalanches in granular materials. *Phys. Rev. Lett.*, **69**(16), 2431–2434.

Brinckmann, S., Kim, J.-Y., and Greer, J. R. (2008). Fundamental differences in mechanical behavior between two types of crystals at the nanoscale. *Phys. Rev. Lett.*, **100**(15), 155502.

Brown, L. (1964). The self-stress of dislocations and the shape of extended nodes. *Philosophical Magazine*, **10**(105), 441–466.

Brugues, A., Anon, E., Conte, V., Veldhuis, J. H., Gupta, M., Colombelli, J., Munoz, J. J., Brodland, G. W., Ladoux, B., and Trepat, X. (2014). Forces driving epithelial wound healing. *Nat. Phys.*, **10**(9), 683–690.

Bryksin, V. and Dorogovtsev, S. (1993*a*). Nonlinear diffusion of magnetic flux in type-ll superconductors. *Zh. Eksp. Teor. Fiz*, **104**, 3735–3758.

Bryksin, V. and Dorogovtsev, S. (1993*b*). Space-time image of the magnetic flux penetrating into type-II superconductors in an applied oscillating magnetic field. *Physica C: Superconductivity*, **215**(1-2), 173–180.

Buchel, A. and Sethna, J. P. (1997). Statistical mechanics of cracks: Fluctuations, breakdown, and asymptotics of elastic theory. *Phys. Rev. E*, **55**, 7669.

Budrikis, Z., Castellanos, D. F., Sandfeld, S., Zaiser, M., and Zapperi, S. (2017). Universal features of amorphous plasticity. *Nature communications*, **8**, 15928.

Budrikis, Z. and Zapperi, S. (2013). Avalanche localization and crossover scaling in amorphous plasticity. *Phys. Rev. E*, **88**(6), 062403.

Buldyrev, S. V., Barabási, A.-L., Caserta, F., Havlin, S., Stanley, H. E., and Vicsek, T. (1992). Anomalous interface roughening in porous media: Experiment and model. *Phys. Rev. A*, **45**(12), R8313–R8316.

Cafiero, R., Gabrielli, A., Marsili, M., and Pietronero, L. (1996). Theory of extremal dynamics with quenched disorder: Invasion percolation and related models. *Phys. Rev. E*, **54**(2), 1406.

Caldarelli, G., Castellano, C., and Petri, A. (1999). Critical behaviour in fracture of disordered media. *Phil. Mag. B*, **79**, 1939.

Caldarelli, G., di Tolla, F., and Petri, A. (1996). Self organization and annealed disorder in fracturing processes. *Phys. Rev. Lett.*, **77**, 2503.

Célarié, F., Prades, S., Bonamy, D., Ferrero, L., Bouchaud, E., Guillot, C., and Marliére, C. (2003). Glass breaks like metal, but at the nanometer scale. *Phys. Rev. Lett.*, **90**, 075504.

Cerruti, B., Durin, G., and Zapperi, S. (2009). Hysteresis and noise in ferromagnetic materials with parallel domain walls. *Phys. Rev. B*, **79**(13), 134429.

Cerruti, B. and Zapperi, S. (2006). Barkhausen noise from zigzag domain walls. *JSTAT*, **2006**(08), P08020.

Chauve, P., Giamarchi, T., and Le Doussal, P. (2000). Creep and depinning in disordered media. *Phys. Rev. B*, **62**(10), 6241–6267.

Chauve, P., Le Doussal, P., and Wiese, K. J. (2001). Renormalization of pinned elastic systems: how does it work beyond one loop. *Phys. Rev. Lett.*, **86**, 1785–1788.

Chen, C.-Y., Yu, F.-C., Lin, S.-C., and Cheung, K.-W. (2007). Discussion of landslide self-organized criticality and the initiation of debris flow. *Earth Surface Processes and Landforms*, **32**(2), 197–209.

Chen, K. and Lin, J. (2010). Investigation of the relationship between primary and secondary shear bands induced by indentation in bulk metallic glasses. *Int. J. Plasticity*, **26**, 1645–1658.

Chen, M. (2008). Mechanical behavior of metallic glasses: microscopic understanding of strength and ductility. *Annu. Rev. Mater. Res.*, **38**, 445–469.

Chen, Y. J., Zapperi, S., and Sethna, J. P. (2015). Crossover behavior in interface depinning. *Phys. Rev. E*, **92**, 022146.

Chepizhko, O., Giampietro, C., Mastrapasqua, E., Nourazar, M., Ascagni, M., Sugni, M., Fascio, U., Leggio, L., Malinverno, C., Scita, G., Santucci, S., Alava, M. J., Zapperi, S., and La Porta, C. A. M. (2016). Bursts of activity in collective cell migration. *PNAS*, **113**(41), 11408–11413.

Chepizhko, O., Lionetti, M. C., Malinverno, C., Giampietro, C., Scita, G., Zapperi, S., and La Porta, C. A. M. (2018). From jamming to collective cell migration through a boundary induced transition. *Soft Matter*, **14**(19), 3774–3782.

Chessa, A., Stanley, H. E., Vespignani, A., and Zapperi, S. (1999). Universality in sandpiles. *Phys. Rev. E*, **59**(1), R12.

Chikkadi, V., Wegdam, G., Bonn, D., Nienhuis, B., and Schall, P. (2011). Long-range strain correlations in sheared colloidal glasses. *Phys. Rev. Lett.*, **107**(19), 198303.

Christian, D. A., Novoselov, K. S., and Geim, A. K. (2006). Barkhausen statistics from a single domain wall in thin films studied with ballistic hall magnetometry. *Phys. Rev. B*, **74**(6), 064403.

Cieplak, M. and Robbins, M. O. (1988). Dynamical transition in quasistatic fluid invasion in porous media. *Phys. Rev. Lett.*, **60**, 2042.

Cizeau, P., Zapperi, S., Durin, G., and Stanley, H. E. (1997). Dynamics of a ferromagnetic domain wall and the Barkhausen effect. *Phys. Rev. Lett.*, **79**, 4669–4672.

Clara-Rahola, J., Brzinski, T., Semwogerere, D., Feitosa, K., Crocker, J., Sato, J., Breedveld, V., and Weeks, E. R. (2015). Affine and nonaffine motions in sheared polydisperse emulsions. *Phys. Rev. E*, **91**(1), 010301.

Clotet, X., Ortín, J., and Santucci, S. (2014). Disorder-induced capillary bursts control intermittency in slow imbibition. *Phys Rev Lett*, **113**(7), 074501.

Colaiori, F. (2008). Exactly solvable model of avalanches dynamics for Barkhausen crackling noise. *Advances in Physics*, **57**(4), 287 – 359.

Colaiori, F., Gabrielli, A., and Zapperi, S. (2002). Rayleigh loops in the random-field Ising model on the Bethe lattice. *Phys. Rev. B*, **65**, 224404.

Corral, A. (2006). Universal earthquake-occurrence jumps, correlations with time, and anomalous diffusion. *Phys. Rev. Lett.*, **97**(17), 178501.

Coslovich, D. and Pastore, G. (2009). Dynamics and energy landscape in a tetrahedral network glass-former: direct comparison with models of fragile liquids. *J. Phys.: Condensed Matter*, **21**(28), 285107.

Costello, R. M., Cruz, K. L., Egnatuk, C., Jacobs, D. T., Krivos, M. C., Louis, T. S., Urban, R. J., and Wagner, H. (2003). Self-organized criticality in a bead pile. *Phys. Rev. E*, **67**(4), 041304.

Cote, P. J. and Meisel, L. V. (1991). Self-organized criticality and the Barkhausen effect. *Phys. Rev. Lett.*, **67**, 1334–1337.

Csikor, F. F., Motz, C., Weygand, D., Zaiser, M., and Zapperi, S. f. (2007). Dislocation avalanches, strain bursts, and the problem of plastic forming at the micrometer sca le. *Science*, **318**(5848), 251–254.

Cule, D. and Hwa, T. (1998). Static and dynamic properties of inhomogeneous elastic media on disordered substrate. *Phys. Rev. B*, **57**, 8235.

Curtin, W. A. (1991). Theory of mechanical properties of ceramic-matrix composites. *J. of the Am. Ceramic Soc.*, **74**(11), 2837–2845.

Curtin, W. A. (1993). The "tough" to brittle transition in brittle matrix composites. *J. Mech. Phys. Solids*, **41**, 217.

da Silveira, R. (1998). Comment on "tricritical behavior in rupture induced by disorder". *Phys. Rev. Lett.*, **80**, 3157.

da Silveira, R. A. and Zapperi, S. (2004). Critical hysteresis from random anisotropy. *Phys. Rev. B*, **69**(21), 212404.

Daguier, P., Nghiem, B., Bouchaud, E., and Creuzet, F. (1997). Pinning and depinning of crack fronts in heterogeneous materials. *Phys. Rev. Lett.*, **78**, 1062.

Dahmen, K. and Sethna, J. P. (1993). Hysteresis loop critical exponents in 6-ε dimensions. *Phys. Rev. Lett.*, **71**, 3222–3225.

Dahmen, K. and Sethna, J. P. (1996). Hysteresis, avalanches, and disorder–induced critical scaling: A renormalization–group approach. *Phys. Rev. B*, **53**, 14872–14905.

Dalton, F. and Corcoran, D. (2001). Self-organized criticality in a sheared granular stick-slip system. *Phys. Rev. E*, **63**, 61312.

Dalton, F. and Corcoran, D. (2002). Basin of attraction of a bounded self-organized critical state. *Phys. Rev. E*, **65**, 31310.

Dalton, F., Farrelly, F., Petri, A., Pietronero, L., Pitolli, L., and Pontuale, G. (2005). Shear stress fluctuations in the granular liquid and solid phases. *Phys. Rev. Lett.*, **95**(13), 138001.

Damento, M. and Demer, L. (1997). Investigation of local events of magnetization reversal in a Nd-Fe-B magnet by use of a hall-effect microprobe. *IEEE Trans. Magn.*, **23**, 1877–1880.

Daniels, H. E. (1945). The statistical theory of strength of bundles of threads. *Proc. R. Soc. London A*, **183**, 405.

Daniels, K. E. (2017). The role of force networks in granular materials. In *EPJ Web of Conferences*, Volume 140, p. 01006. EDP Sciences.

Daniels, K. E. and Hayman, N. W. (2008). Force chains in seismogenic faults visualized with photoelastic granular shear experiments. *Journal of Geophysical Research: Solid Earth*, **113**(B11).

Daniels, K. E., Kollmer, J. E., and Puckett, J. G. (2017). Photoelastic force measurements in granular materials. *Review of Scientific Instruments*, **88**(5), 051808.

Dante, L., Durin, G., Magni, A., , and Zapperi, S. (2002). Low-field hysteresis in disordered ferromagnets. *Phys. Rev. B*, **65**, 144441.

de Arcangelis, L., Redner, S., and Herrmann, H. J. (1985). A random fuse model for breaking processes. *J. Phys. (Paris) Letters*, **46**(13), 585–590.

de Gennes, P. G. (1966). *Superconductivity in metals and alloys*. Benjamin New

York.

de Gennes, P. G. (1999). Granular matter: a tentative view. *Rev. Mod. Phys.*, **71**, S374.

de Queiroz, S. L. A. and Bahiana, M. (2001). Finite driving rates in interface models of Barkhausen noise. *Phys. Rev. E*, **64**, 066127.

Delaplace, A., Schmittbuhl, J., and Maloy, K. J. (1999). High resolution description of a crack front in a heterogeneous plexiglas block. *Phys. Rev. E*, **60**, 1337.

Demkowicz, M. J. and Argon, A. S. (2005). Autocatalytic avalanches of unit inelastic shearing events are the mechanism of plastic deformation in amorphous silicon. *Phys. Rev. B*, **72**(24), 245206.

Dennin, M. (2004). Statistics of bubble rearrangements in a slowly sheared two-dimensional foam. *Phys. Rev. E*, **70**(4), 041406.

Derrida, B. (1987). Dynamical phase transition in nonsymmetric spin glasses. *J. Phys. A Math. Gen.*, **20**(11), L721.

Detcheverry, F., Kierlik, E., Rosinberg, M. L., and Tarjus, G. (2003). Local mean-field study of capillary condensation in silica aerogels. *Phys. Rev. E*, **68**(6), 061504.

Detcheverry, F., Kierlik, E., Rosinberg, M. L., and Tarjus, G. (2005). Helium condensation in aerogel: Avalanches and disorder-induced phase transition. *Phys. Rev. E* (5), 051506.

Devincre, B., Hoc, T., and Kubin, L. (2008). Dislocation mean free paths and strain hardening of crystals. *Science*, **320**(5884), 1745–1748.

DeWit, G. and Koehler, J. (1959). Interaction of dislocations with an applied stress in anisotropic crystals. *Phys. Rev.*, **116**(5), 1113.

Dhar, D. (1990). Self-organized critical state of sandpile automaton models. *Phys. Rev. Lett.*, **64**(14), 1613.

Dhar, D. (1999). The Abelian sandpile and related models. *Physica A: Statistical Mechanics and its Applications*, **263**(1-4), 4–25.

Dhar, D. (2006). Theoretical studies of self-organized criticality. *Physica A: Statistical Mechanics and its Applications*, **369**(1), 29–70.

Dhar, D. and Majumdar, S. (1990). Abelian sandpile model on the Bethe lattice. *J. Phys. A: Mathematical and General*, **23**(19), 4333.

Dhar, D. and Ramaswamy, R. (1989). Exactly solved model of self-organized critical phenomena. *Phys. Rev. Lett.*, **63**(16), 1659.

Dickman, R. and Campelo, J. (2003). Avalanche exponents and corrections to scaling for a stochastic sandpile. *Phys. Rev. E*, **67**(6), 066111.

Dickman, R., Munoz, M. A., Vespignani, A., and Zapperi, S. (2000). Paths to self-organized criticality. *Braz. J. Phys.*, **30**, 27–39.

Dickman, R., Vespignani, A., and Zapperi, S. (1998). Self-organized criticality as an absorbing-state phase transition. *Phys. Rev. E*, **57**(5), 5095.

Dimiduk, D. M., Woodward, C., LeSar, R., and Uchic, M. D. (2006). Scale-Free Intermittent Flow in Crystal Plasticity. *Science*, **312**, 1188–1190.

Dobrinevski, A., Le Doussal, P., and Wiese, K. J. (2015). Avalanche shape and exponents beyond mean-field theory. *Europhys. Lett.*, **108**(6), 66002.

Dougherty, A. and Carle, N. (1998). Distribution of avalanches in interfacial motion in a porous medium. *Phys. Rev. E*, **58**(3), 2889–2893.

Drossel, B. and Dahmen, K. (1998). Depinning of a domain wall in the 2d random-field Ising model. *Eur. Phys. J. B*, **3**, 485–496.

Durin, G., Magni, A., and Bertotti, G. (1995*a*). Fractal properties of the Barkhausen effect. *J. Magn. Magn. Mat.*, **140-144**, 1835–1836.

Durin, G., Magni, A., and Bertotti, G. (1995*b*). Fractals, scaling and the question of self-organized criticality in magnetization processes. *Fractals*, **3**, 351.

Durin, G. and Zapperi, S. (2000). Scaling exponents for Barkhausen avalanches in polycrystalline and amorphous ferromagnets. *Phys. Rev. Lett.*, **84**, 4075–4078.

Durin, G. and Zapperi, S. (2002). On the power spectrum of magnetization noise. *J. Magn. Magn. Mat.*, **242-245**, 1085–1088.

Durin, G. and Zapperi, S. (2005). *The Science of Hysteresis: Physical Modeling, Micromagnetics, and Magnetization Dynamics*, Volume II, Chapter III "The Barkhausen noise", pp. 181–267. Academic Press, Amsterdam. (Preprint on cond-mat/0404512).

Durin, G. and Zapperi, S. (2006). The role of stationarity in magnetic crackling noise. *JSTAT*, **2006**(01), P01002.

Dussauge, C., Grasso, J.-R., and Helmstetter, A. (2003). Statistical analysis of rock-fall volume distributions: Implications for rockfall dynamics. *Journal of Geophysical Research: Solid Earth*, **108**(B6), 2286.

Edwards, S. F. and Anderson, P. W. (1975). Theory of spin glasses. *J. Phys. F: Metal Physics*, **5**(5), 965.

Enomoto, Y. (1994*a*). Bloch wall motion in random inpurities. *J. Magn. Magn. Mat.*, **129**, L146–150.

Enomoto, Y. (1994*b*). Magnetic domain wall depinning in a disordered medium. *J. Magn. Magn. Mat.*, **138**, L1–5.

Enomoto, Y. (1997). Crossover behavior of magnetization noise properties in disordered ferromagnets. *J. Magn. Magn. Mat.*, **174**, 155–159.

Ertas, D. and Kardar, M. (1994). Critical dynamics of contact line depinning. *Phys. Rev. E*, **49**, R2532–2535.

Eshelby, J. D. (1957). The determination of the elastic field of an ellipsoidal inclusion, and related problems. *Proc. Royal Soc. of London. Series A. Mathematical and Physical Sciences*, **241**(1226), 376–396.

Evesque, P. and Rajchenbach, J. (1989). Instability in a sand heap. *Phys. Rev. Lett.*, **62**(1), 44–46.

Ewald, P. P. (1921). Die Berechnung optischer und elektrostatischer Gitterpotentiale. *Ann. Phys.*, **369**, 253–287.

Fabeny, B. and Curtin, W. A. (1996). Damage-enhanced creep and rupture in fiber-reinforced composites. *Acta Materialia*, **44**, 3439.

Faillettaz, J., Louchet, F., and Grasso, J.-R. (2004). Two-threshold model for scaling laws of noninteracting snow avalanches. *Phys. Rev. Lett.*, **93**(20), 208001.

Falk, M. L. and Langer, J. S. (1998). Dynamics of viscoplastic deformation in amorphous solids. *Phys. Rev. E*, **57**(6), 7192.

Feng, S. and Sen, P. N. (1984). Percolation on elastic networks: New exponent and threshold. *Phys. Rev. Lett.*, **52**(3), 216–219.

Field, S., Witt, J., Nori, F., and Ling, X. (1995). Superconducting vortex avalanches.

Phys. Rev. Lett., **74**, 1206.

Fisher, D. S. (1985). Sliding charge-density waves as a dynamic critical phenomenon. *Phys. Rev. B*, **31**, 1396–1427.

Floyd, C., Levine, H., Jarzynski, C., and Papoian, G. A. (2021). Understanding cytoskeletal avalanches using mechanical stability analysis. *PNAS*, **118**(41), e2110239118.

Flyvbjerg, H., Sneppen, K., and Bak, P. (1993). Mean field theory for a simple model of evolution. *Phys. Rev. Lett.*, **71**(24), 4087.

Font-Clos, F., Zapperi, S., and La Porta, C. A. (2018). Topography of epithelial–mesenchymal plasticity. *PNAS*, **115**(23), 5902–5907.

Font-Clos, F., Zapperi, S., and La Porta, C. A. (2021). Classification of triple-negative breast cancers through a boolean network model of the epithelial-mesenchymal transition. *Cell Systems*, **12**(5), 457–462.

Foreman, A. J. E. (1967). The bowing of a dislocation segment. *Phil. Mag.*, **15**, 1011.

Frette, V., Christensen, K., Malthe-Sorenssen, M., Feder, J., Jossang, T., and Meakin, P. (1996). Avalanche dynamics in a pile of rice. *Nature*, **379**, 49.

Friedel, J. (1967). *Dislocations*. Pergamon Press, Oxford.

Friedl, P. and Gilmour, D. (2009). Collective cell migration in morphogenesis, regeneration and cancer. *Nat. Rev. Mol. Cell Biol.*, **10**(7), 445–57.

Furuberg, L., Feder, J., Aharony, A., and Jøssang, T. (1988). Dynamics of invasion percolation. *Phys. Rev. Lett.*, **61**(18), 2117.

Furuberg, L., Måløy, K. J., and Feder, J. (1996). Intermittent behavior in slow drainage. *Phys. Rev. E*, **53**(1), 966–977.

Gabrielov, A. (1993). Abelian avalanches and Tutte polynomials. *Physica A: Statistical Mechanics and its Applications*, **195**(1-2), 253–274.

Gao, H. and Rice, J. R. (1989). A first-order perturbation analysis of crack trapping by arrays of obstacles. *J. Appl. Mech.*, **56**, 828.

Garcimartin, A., Guarino, A., Bellon, L., and Ciliberto, S. (1997). Statistical properties of fracture precursors. *Phys. Rev. Lett.*, **79**(17), 3202.

Géminard, J.-C., Losert, W., and Gollub, J. P. (1999). Frictional mechanics of wet granular material. *Phys. Rev. E*, **59**(5), 5881–5890.

Geng, J. and Behringer, R. P. (2005). Slow drag in two-dimensional granular media. *Phys. Rev. E*, **71**, 11302.

George, J. T., Jolly, M. K., Xu, S., Somarelli, J. A., and Levine, H. (2017). Survival outcomes in cancer patients predicted by a partial EMT gene expression scoring metric. *Cancer Res*, **77**(22), 6415–6428.

Giamarchi, T. (2009). *Disordered Elastic Media*, pp. 2019–2038. Springer New York, New York, NY.

Giamarchi, T. and Le Doussal, P. (1996). Moving glass phase of driven lattices. *Phys. Rev. Lett.*, **76**(18), 3408–3411.

Gilchrist, J. and van der Beek, C. J. (1994). Nonlinear diffusion in hard and soft superconductors. *Physica C*, **231**, 147.

Giusti, C., Papadopoulos, L., Owens, E. T., Daniels, K. E., and Bassett, D. S. (2016). Topological and geometric measurements of force-chain structure. *Phys. Rev. E*, **94**(3), 032909.

Golubovic, L. and Feng, S. (1991). Rate of microcrack nucleation. *Phys. Rev. A*, **43**, 5223.

Golubovic, L. and Pedrera, A. (1995). Mechanism of time-delayed fractures. *Phys. Rev. E*, **51**, 2799.

Gould, S. J. and Eldredge, N. (1977). Punctuated equilibria: the tempo and mode of evolution reconsidered. *Paleobiology*, **3**(2), 115–151.

Gov, N. S. (2014). *Cell and matrix mechanics*, pp. 219–238. CRC Press.

Grassberger, P. and de la Torre, A. (1979). Reggeon field theory (Schlögl's first model) on a lattice: Monte Carlo calculations of critical behaviour. *Annals of Physics*, **122**(2), 373–396.

Grassberger, P. and Manna, S. (1990). Some more sandpiles. *J. Phys.(France)*, **51**, 1077.

Guarino, A., Ciliberto, S., Garcimartin, A., Zei, M., and Scorretti, R. (2002). Failure time and critical behaviour of fracture precursors in heterogeneous materials. *Eur. Phys. J. B*, **26**, 141.

Guarino, A., Garcimartin, A., and Ciliberto, S. (1998). An experimental test of the critical behaviour of fracture precursors. *Eur. Phys. J. B*, **6**(1), 13–24.

Gutenberg, B. (1954). *Seismicity of the Earth and Associated Phenomena*. Princeton University Press, Princeton.

Gutfreund, H., Reger, J. D., and Young, A. P. (1988). The nature of attractors in an asymmetric spin glass with deterministic dynamics. *J. Phys. A Math. Gen.*, **21**(12), 2775.

Guyer, R. A. and McCall, K. R. (1996). Capillary condensation, invasion percolation, hysteresis, and discrete memory. *Phys. Rev. B*, **54**, 18–21.

Haeger, A., Krause, M., Wolf, K., and Friedl, P. (2014). Cell jamming: Collective invasion of mesenchymal tumor cells imposed by tissue confinement. *Biochim. Biophys. Acta*, **1840**(8), 2386–95.

Hansen, A. and Hemmer, P. C. (1994). Burst avalanches in bundles of fibers: Local versus global load-sharing. *Phys. Lett. A*, **184**, 394.

Hansen, A., Hinrichsen, E. L., and Roux, S. (1991). Scale invariant disorder in fracture and related breakdown phenomena. *Phys. Rev. B*, **43**, 665.

Hansen, A., Roux, S., and Herrmann, H. J. (1989). Rupture of central-force lattices. *J. Physique*, **50**, 733–744.

Haraguchi, S., Fukuda, Y., Sugisaki, Y., and Yamanaka, N. (1999). Pulmonary carcinosarcoma: Immunohistochemical and ultrastructural studies. *Pathol Int*, **49**(10), 903–8.

Harlow, D. G. and Phoenix, S. L. (1978). The chain-of-bundles probability model for the strength of fibrous materials I: Analysis and conjectures. *J. Composite Mater.*, **12**, 195.

Harris, T. E. (1963). *The Theory of Branching Processes*. Volume 6. Springer Berlin.

Hassold, G. N. and Srolovitz, D. J. (1989). Brittle fracture in materials with random defects. *Phys. Rev. B*, **39**, 9273–9281.

Hébraud, P., Lequeux, F., Munch, J., and Pine, D. (1997). Yielding and rearrangements in disordered emulsions. *Phys. Rev. Lett.*, **78**(24), 4657.

Heiden, C. and Rochlin, G. (1968). Flux jump size distribution in low-κ type-II superconductors. *Phys. Rev. Lett.*, **21**(10), 691.

Held, G. A., Solina, D. H., Solina, H., Keane, D. T., Haag, W. J., Horn, P. M., and Grinstein, G. (1990). Experimental study of critical-mass fluctuations in an evolving sandpile. *Phys. Rev. Lett.*, **65**(9), 1120–1123.

Hemmer, P. C. and Hansen, A. (1992). The distribution of simultaneous fiber failures in fiber bundles. *J. Appl. Mech.*, **59**, 909.

Hentschel, H. G. E., Jaiswal, P. K., Procaccia, I., and Sastry, S. (2015). Stochastic approach to plasticity and yield in amorphous solids. *Phys. Rev. E*, **92**, 062302.

Hergarten, S. (2002). *Self organized criticality in earth systems*. Volume 2. Springer.

Herrmann, H. J., Hovi, J.-P., and Luding, S. (2013). *Physics of dry granular media*. Volume 350. Springer Science & Business Media.

Herrmann, H. J. and Roux, S. (1990). *Statistical Models for the Fracture of Disordered Media*. North-Holland, Amsterdam.

Hidalgo, R. C., Kun, F., and Herrmann, H. J. (2002). Creep rupture of viscoelastic fiber bundles. *Phys. Rev. E*, **65**, 032502.

Hilzinger, H.-R. (1976). Computer simulation of magnetic domain wall pinning. *Phys. Status Solidi A*, **38**, 487–496.

Hilzinger, H. R. (1977). Scaling relations in magnetic, mechanical and superconducting pinning theory. *Philos. Mag.*, **36**, 225–234.

Hilzinger, H. R. and Kronmüller, H. (1976). Statistical theory of the pinning field of Bloch walls by randomly distributed defects. *J. Magn. Magn. Mat.*, **2**, 11–17.

Hinrichsen, H. (2000). Non-equilibrium critical phenomena and phase transitions into absorbing states. *Advances in physics*, **49**(7), 815–958.

Hirth, J. P. and Lothe, J. (1992). *Theory of Dislocations*. Krieger Publishing Company.

Horbach, J., Kob, W., Binder, K., and Angell, C. A. (1996). Finite size effects in simulations of glass dynamics. *Phys. Rev. E*, **54**(6), R5897.

Hoshen, J., Stauffer, D., Bishop, G. H., Harrison, R. J., and Quinn, G. D. (1979). Monte Carlo experiments on cluster size distribution in percolation. *J. Phys. A: Mathematical and General*, **12**(8), 1285.

Howell, D., Behringer, R. P., and Veje, C. (1999). Stress fluctuations in a 2d granular Couette experiment: A continuous transition. *Phys. Rev. Lett.*, **82**(26), 5241–5244.

Huang, B., Lu, M., Jia, D., Ben-Jacob, E., Levine, H., and Onuchic, J. N. (2017). Interrogating the topological robustness of gene regulatory circuits by randomization. *PLOS Computational Biology*, **13**(3), 1–21.

Huang, P. Y., Kurasch, S., Alden, J. S., Shekhawat, A., Alemi, A. A., McEuen, P. L., Sethna, J. P., Kaiser, U., and Muller, D. A. (2013). Imaging atomic rearrangements in two-dimensional silica glass: watching silica's dance. *science*, **342**(6155), 224–227.

Huang, S. (2012). The molecular and mathematical basis of Waddington's epigenetic landscape: A framework for post-Darwinian biology? *Bioessays*, **34**(2), 149–57.

Huang, S., Eichler, G., Bar-Yam, Y., and Ingber, D. E. (2005). Cell fates as high-dimensional attractor states of a complex gene regulatory network. *Phys. Rev. Lett.*, **94**, 128701.

Huang, S., Guo, Y.-P., May, G., and Enver, T. (2007). Bifurcation dynamics in lineage-commitment in bipotent progenitor cells. *Dev Biol*, **305**(2), 695–713.

Hubert, A. and Schäfer, R. (1998). *Magnetic domains*. Springer, New York.

Hughes, T. R., Marton, M. J., Jones, A. R., Roberts, C. J., Stoughton, R., Armour, C. D., Bennett, H. A., Coffey, E., Dai, H., He, Y. D. et al. (2000). Functional discovery via a compendium of expression profiles. *Cell*, **102**(1), 109–126.

Ilina, O. and Friedl, P. (2009). Mechanisms of collective cell migration at a glance. *J Cell Sci*, **122**(Pt 18), 3203–8.

Imry, Y. and Ma, S.-k. (1975). Random-field instability of the ordered state of continuous symmetry. *Phys. Rev. Lett.*, **35**(21), 1399–1401.

Ispánovity, P. D., Laurson, L., Zaiser, M., Groma, I., Zapperi, S., and Alava, M. J. (2014). Avalanches in 2d dislocation systems: Plastic yielding is not depinning. *Phys. Rev. Lett.*, **112**(23), 235501.

Ivashkevich, E. V., Ktitarev, D. V., and Priezzhev, V. B. (1994). Waves of topplings in an Abelian sandpile. *Physica A: Statistical Mechanics and its Applications*, **209**(3-4), 347–360.

Jaeger, H. M., Liu, C., Nagel, S. R., and Witten, T. A. (1990). Friction in granular flows. *Europhys.Lett.*, **11**(7), 619–624.

Jaeger, H. M., Liu, C.-h., and Nagel, S. R. (1989). Relaxation at the angle of repose. *Phys. Rev. Lett.*, **62**(1), 40–43.

Jaeger, H. M., Nagel, S. R., and Behringer, R. P. (1996). Granular solids, liquids, and gases. *Rev. Mod. Phys.*, **68**(4), 1259–1273.

Janićević, S., Laurson, L., Måløy, K. J., Santucci, S., and Alava, M. J. (2016). Interevent correlations from avalanches hiding below the detection threshold. *Phys. Rev. Lett.*, **117**(23), 230601.

Janićević, S., Ovaska, M., Alava, M. J., and Laurson, L. (2015). Avalanches in 2d dislocation systems without applied stresses. *JSTAT*, **2015**(7), P07016.

Janssen, H. A. (1895). Versuche über Getreidedruck in Silozellen. *Zeitschr. d. Vereines deutscher Ingenieure*, **39**(35), 1045–1049.

Jensen, H. J., Christensen, K., and Fogedby, H. C. (1989). 1/f noise, distribution of lifetimes, and a pile of sand. *Phys. Rev. B*, **40**(10), 7425.

Ji, H. and Robbins, M. (1991). Transition from compact to self-similar growth in disordered systems: Fluid invasion and magnetic-domain growth. *Phys. Rev. A*, **44**, 2538–2542.

Ji, H. and Robbins, M. (1992). Percolative, self-affine, and faceted domain growth in random three-dimensional magnets. *Phys. Rev. B*, **46**, 14519.

Jolly, M. K., Huang, B., Lu, M., Mani, S. A., Levine, H., and Ben-Jacob, E. (2014). Towards elucidating the connection between epithelial-mesenchymal transitions and stemness. *J. Royal Soc. Interface*, **11**(101), 20140962.

Jolly, M. K., Tripathi, S. C., Jia, D., Mooney, S. M., Celiktas, M., Hanash, S. M., Mani, S. A., Pienta, K. J., Ben-Jacob, E., and Levine, H. (2016). Stability of the hybrid epithelial/mesenchymal phenotype. *Oncotarget*, **7**(19), 27067–84.

Kahng, B., Batrouni, G. G., Redner, S., de Arcangelis, L., and Herrmann, H. J. (1988). Electrical breakdown in a fuse network with random, continuously distributed breaking strengths. *Phys. Rev. B*, **37**(13), 7625–7637.

Kardar, M., Parisi, G., and Zhang, Y.-C. (1986). Dynamic scaling of growing interfaces. *Phys. Rev. Lett.*, **56**, 889–892.

Karmakar, S., Lerner, E., and Procaccia, I. (2010). Statistical physics of the yielding transition in amorphous solids. *Phys. Rev. E*, **82**, 055103.

Katz, A. J., Thompson, A. H., and Raschke, R. A. (1988). Numerical simulation of resistance steps for mercury injection under the influence of gravity. *Phys. Rev. A*, **38**(9), 4901–4904.

Kauffman, S. A. (1969). Metabolic stability and epigenesis in randomly constructed genetic nets. *J. Theor. Biol.*, **22**(3), 437–67.

Kauffman, S. A. (1993). *The Origins of Order: Self-Organization and Selection in Evolution*. Oxford University Press, USA.

Kegel, W. and van Blaaderen, A. (2000). Direct observation of dynamical heterogeneities in colloidal hard-sphere suspensions. *Science*, **287**, 290.

Kertész, J. and Kiss, L. (1990). The noise spectrum in the model of self-organised criticality. *J. Phys. A: Mathematical and General*, **23**(9), L433.

Khalil, A. A. and Friedl, P. (2010). Determinants of leader cells in collective cell migration. *Integr Biol (Camb)*, **2**(11-12), 568–74.

Kierlik, E., Monson, P. A., Rosinberg, M. L., Sarkisov, L., and Tarjus, G. (2001). Capillary condensation in disordered porous materials: Hysteresis versus equilibrium behavior. *Phys. Rev. Lett.*, **87**(5), 055701.

Kim, D.-H., Choe, S.-B., and Shin, S.-C. (2003*a*). Direct observation of Barkhausen avalanche in co thin films. *Phys. Rev. Lett.*, **90**, 087203.

Kim, D.-H., Choe, S.-B., and Shin, S.-C. (2003*b*). Time-resolved observation of Barkhausen avalanche in co thin films using magneto-optical microscope magnetometer. *J. Appl. Phys.*, **93**, 6564–6566.

Kirchner, J. W. and Weil, A. (1998). No fractals in fossil extinction statistics. *Nature*, **395**(6700), 337–338.

Kirkpatrick, S. (1973). Percolation and conduction. *Rev. Mod. Phys.*, **45**(4), 574.

Klaus, A., Yu, S., and Plenz, D. (2011). Statistical analyses support power law distributions found in neuronal avalanches. *PloS one*, **6**(5), e19779.

Klimas, A. J., Valdivia, J., Vassiliadis, D., Baker, D., Hesse, M., and Takalo, J. (2000). Self-organized criticality in the substorm phenomenon and its relation to localized reconnection in the magnetospheric plasma sheet. *Journal of Geophysical Research: Space Physics*, **105**(A8), 18765–18780.

Kloster, P. M., Hemmer, C., and Hansen, A. (1997). Burst avalanches in solvable models of fibrous materials. *Phys. Rev. E*, **56**, 2615.

Koch, T. M., Münster, S., Bonakdar, N., Butler, J. P., and Fabry, B. (2012). 3d traction forces in cancer cell invasion. *PLoS One*, **7**(3), e33476.

Koiller, B., Ji, H., and Robbins, M. (1992). Effect of disorder and lattice type on domain-wall motion in two dimensions. *Phys. Rev. B*, **46**, 5258.

Koiller, B. and Robbins, M. O. (2000). Morphology transitions in three-dimensional domain growth with Gaussian random fields. *Phys. Rev. B*, **62**, 5771–5778.

Koplik, J. and Levine, H. (1985). Interface moving through a random background. *Phys. Rev. B*, **32**, 280.

Krajcinovic, D. and Silva, M. A. G. (1982). Statistical aspects of the continuous damage theory. *Int. J. Solids Structures*, **18**, 551.

Kun, F., Hidalgo, R. C., Herrmann, H. J., and Pál, K. F. (2003). Scaling laws of creep rupture of fiber bundles. *Phys. Rev. E*, **67**, 061802.

Kun, F., Zapperi, S., and Herrmann, H. J. (2000). Damage in fiber bundle models. *Eur. Phys. J. B*, **17**, 269.

Kuntz, M. C. and Sethna, J. P. (2000). Noise in disordered systems: The power spectrum and dynamic exponents in avalanche models. *Phys. Rev. B*, **62**, 11699–11708.

La Porta, C. A. and Zapperi, S. (2019). *Cell Migrations: Causes and Functions.* Volume 1146. Springer.

La Porta, C. A. and Zapperi, S. (2020). Phase transitions in cell migration. *Nature Reviews Physics*, **2**(10), 516–517.

La Porta, C. A. M. and Zapperi, S. (2017). *The Physics of Cancer.* Cambridge University Press.

Labusch, R. (1969). Calculation of the critical field gradient in type-II superconductors. *Cryst. Lattice Defects*, **1**, 1.

Labusch, R. (1970). A statistical theory of solid solution hardening. *physica status solidi (b)*, **41**(2), 659–669.

Lacombe, F., Herrmann, H. J., and Zapperi, S. (2000). Dilatancy and friction in granular media. *Eur. Phys. J. E*, **2**, 181.

Lange, J. R. and Fabry, B. (2013). Cell and tissue mechanics in cell migration. *Exp Cell Res*, **319**(16), 2418–23.

Larkin, A. I. (1970). Effect of Inhomogeneties on the Structure of the Mixed State of Superconductors. *Soviet J. Exp. Theor. Phys.*, **31**, 784.

Lauritsen, K. B., Zapperi, S., and Stanley, H. E. (1996). Self-organized branching processes: Avalanche models with dissipation. *Phys. Rev. E*, **54**(3), 2483.

Laurson, L., Alava, M. J., and Zapperi, S. (2005). Power spectra of self-organized critical sandpiles. *JSTAT*, **2005**(11), L11001.

Laurson, L., Illa, X., and Alava, M. J. (2009). The effect of thresholding on temporal avalanche statistics. *J. Stat. Mech.*, P01019.

Laurson, L., Santucci, S., and Zapperi, S. (2010). Avalanches and clusters in planar crack front propagation. *Phys. Rev. E*, **81**(4), 046116.

Le Doussal, P. and Giamarchi, T. (1998). Moving glass theory of driven lattices with disorder. *Phys. Rev. B*, **57**, 11356–11403.

Le Doussal, P., Middleton, A. A., and Wiese, K. J. (2009). Statistics of static avalanches in a random pinning landscape. *Phys. Rev. E*, **79**(5), 050101.

Le Doussal, P., Wiese, K., and Chauve, P. (2002). Two-loop functional renormalization group theory of the depinning transition. *Phys. Rev. B*, **66**, 174201.

Le Doussal, P. and Wiese, K. J. (2012). Distribution of velocities in an avalanche. *Europhys. Lett.*, **97**(4), 46004.

Le Doussal, P. and Wiese, K. J. (2012). First-principles derivation of static avalanche-size distributions. *Phys. Rev. E*, **85**, 061102.

Le Doussal, P. and Wiese, K. J. (2013). Avalanche dynamics of elastic interfaces. *Phys. Rev. E*, **88**(2), 022106.

Le Doussal, P. and Wiese, K. J. (2015). Exact mapping of the stochastic field theory for Manna sandpiles to interfaces in random media. *Phys. Rev. Lett.*, **114**(11), 110601.

Le Doussal, Pierre, P., Müller, M., and Wiese, K. J. (2010). Avalanches in mean-field models and the Barkhausen noise in spin-glasses. *Europhys. Lett.*, **91**(5), 57004.

Le Priol, C., Le Doussal, P., and Rosso, A. (2021). Spatial clustering of depinning avalanches in presence of long-range interactions. *Phys. Rev. Lett.*, **126**(2), 025702.

Leath, P. L. and Duxbury, P. M. (1994). Fracture of heterogeneous materials with continuous distributions of local breaking strengths. *Phys. Rev. B*, **49**, 14905.

Lee, E., Salic, A., Krüger, R., Heinrich, R., Kirschner, M. W., and Nusse, R. (2003). The roles of apc and axin derived from experimental and theoretical analysis of the wnt pathway. *PLoS biology*, **1**(1), e10.

Lehtinen, A., Costantini, G., Alava, M. J., Zapperi, S., and Laurson, L. (2016). Glassy features of crystal plasticity. *Phys. Rev. B*, **94**(6), 064101.

Lemerle, S., Ferré, J., Chappert, C., Mathet, V., Giamarchi, T., and Le Doussal, P. (1998). Domain wall creep in ultrathin films. *Phys. Rev. Lett.*, **80**, 849–852.

Lenormand, R. and Zarcone, C. (1985). Invasion percolation in an etched network: Measurement of a fractal dimension. *Phys. Rev. Lett.*, **54**, 2226–2229.

Leschhorn, H. (1993). Interface depinning in a disordered medium – numerical results. *Physica A: Statistical and Theoretical Physics*, **195**(3-4), 324 – 335.

Leschhorn, H. (1996). Anisotropic interface depinning: Numerical results. *Phys. Rev. E*, **54**(2), 1313–1320.

Leschhorn, H., Nattermann, T., Stepanow, S., and Tang, L. H. (1997). Driven interface depinning in a disordered medium. *Ann. Physik*, **6**, 1–34.

Li, C. and Wang, J. (2013). Quantifying Waddington landscapes and paths of non-adiabatic cell fate decisions for differentiation, reprogramming and transdifferentiation. *J R Soc Interface*, **10**(89), 20130787.

Li, F., Long, T., Lu, Y., Ouyang, Q., and Tang, C. (2004). The yeast cell-cycle network is robustly designed. *PNAS*, **101**(14), 4781–4786.

Lieneweg, U. and Grosse-Nobis, W. (1972). Spatial extent of the flux due to local irreversible changes of the magnetic polarization in ferromagnetic strips and rods. *Inter. J. Magnetism*, **3**, 11–16.

Lilly, M. P., Finley, P. T., and Hallock, R. B. (1993). Memory, congruence, and avalanche events in hysteretic capillary condensation. *Phys. Rev. Lett.*, **71**(25), 4186–4189.

Lilly, M. P. and Hallock, R. B. (2001). Avalanche behavior in the draining of superfluid helium from the porous material nuclepore. *Phys. Rev. B*, **64**(2), 024516.

Lilly, M. P., Wootters, A. H., and Hallock, R. B. (1996). Spatially extended avalanches in a hysteretic capillary condensation system: Superfluid ^{4}He in nuclepore. *Phys. Rev. Lett.*, **77**(20), 4222–4225.

Lilly, M. P., Wootters, A. H., and Hallock, R. B. (2002). Avalanches in the draining of nanoporous nuclepore mediated by the superfluid helium film. *Phys. Rev. B*, **65**(10), 104503.

Lin, J., Lerner, E., Rosso, A., and Wyart, M. (2014a). Scaling description of the yielding transition in soft amorphous solids at zero temperature. *PNAS*, **111**(40), 14382–14387.

Lin, J., Saade, A., Lerner, E., Rosso, A., and Wyart, M. (2014b). On the density of shear transformations in amorphous solids. *Europhys. Lett.*, **105**(2), 26003.

Liu, A. J. and Nagel, S. R. (1998). Nonlinear dynamics: Jamming is not just cool any more. *Nature*, **396**, 21.

Liu, A. J., Nagel, S. R., and Langer, J. S. (2010). The jamming transition and the marginally jammed solid. *Annu. Rev. Condens. Matter Phys.*, **1**, 347–369.

Liu, C., Ferrero, E. E., Puosi, F., Barrat, J.-L., and Martens, K. (2016). Driving rate dependence of avalanche statistics and shapes at the yielding transition. *Phys. Rev. Lett.*, **116**(6), 065501.

Liu, C., Nagel, S. R., Schecter, D. A., Coppersmith, S. N., Majumdar, S., Narayan, O., and Witten, T. A. (1995). Force fluctuations in bead packs. *Science*, **269**, 513.

Lopez, J. M. (1999). Scaling approach to calculate critical exponents in anomalous surface roughening. *Phys. Rev. Lett.*, **83**, 4594.

Losert, W., Géminard, J.-C., Nasuno, S., and Gollub, J. P. (2000). Mechanisms for slow strengthening in granular materials. *Phys. Rev. E*, **61**(4), 4060–4068.

Lübeck, S. (2000). Moment analysis of the probability distribution of different sand-pile models. *Phys. Rev. E*, **61**(1), 204.

Lübeck, S. (2004). Universal scaling behavior of non-equilibrium phase transitions. *International Journal of Modern Physics B*, **18**(31n32), 3977–4118.

Lübeck, S. and Heger, P. (2003). Universal finite-size scaling behavior and universal dynamical scaling behavior of absorbing phase transitions with a conserved field. *Phys. Rev. E*, **68**(5), 056102.

Ma, S., Kemmeren, P., Gresham, D., and Statnikov, A. (2014). De-novo learning of genome-scale regulatory networks in S. cerevisiae. *Plos one*, **9**(9), e106479.

Maes, C., van Moffaert, A., Frederix, H., and Strauven, H. (1998). Criticality in creep experiments on cellular glass. *Phys. Rev. B*, **57**, 4987.

Magni, A., Durin, G., Zapperi, S., and Sethna, J. P. (2009). Visualization of avalanches in magnetic thin films: Temporal processing. *JSTAT*, **2009**(01), P01020 (10pp).

Majumdar, S. N. and Dhar, D. (1991). Height correlations in the Abelian sandpile model. *J. Phys. A: Mathematical and General*, **24**(7), L357.

Majumdar, S. N. and Dhar, D. (1992). Equivalence between the Abelian sandpile model and the $q \to 0$ limit of the Potts model. *Physica A: Statistical Mechanics and its Applications*, **185**(1-4), 129–145.

Malinverno, C., Corallino, S., Giavazzi, F., Bergert, M., Li, Q., Leoni, M., Disanza, A., Frittoli, E., Oldani, A., Martini, E., Lendenmann, T., Deflorian, G., Beznoussenko, G. V., Poulikakos, D., Haur, O. K., Uroz, M., Trepat, X., Parazzoli, D., Maiuri, P., Yu, W., Ferrari, A., Cerbino, R., and Scita, G. (2017). Endocytic reawakening of motility in jammed epithelia. *Nat. Mater.*, **16**(5), 587–596.

Maloney, C. and Lemaitre, A. (2004). Subextensive scaling in the athermal, quasistatic limit of amorphous matter in plastic shear flow. *Phys. Rev. Lett.*, **93**(1), 016001.

Maloney, C. and Robbins, M. (2009). Anisotropic power law strain correlations in sheared amorphous 2d solids. *Phys. Rev. Lett.*, **102**(22), 225502.

Maloney, C. E. and Lemaître, A. (2006). Amorphous systems in athermal, quasistatic shear. *Phys. Rev. E*, **74**(1), 016118.

Måløy, K. J., Furuberg, L., Feder, J., and Jøssang, T. (1992). Dynamics of slow drainage in porous media. *Phys. Rev. Lett.*, **68**(14), 2161.

Maloy, K. J., Santucci, S., Schmittbuhl, J., and Toussaint, R. (2006). Local waiting time fluctuations along a randomly pinned crack front. *Phys. Rev. Lett.*, **96**, 045501.

Maloy, K. J. and Schmittbuhl, J. (2001). Dynamical event during slow crack propagation. *Phys. Rev. Lett.*, **87**, 105502.

Manna, S. (1990). Large-scale simulation of avalanche cluster distribution in sand pile model. *Journal of statistical physics*, **59**(1), 509–521.

Manna, S. (1991). Two-state model of self-organized criticality. *J. Phys. A: Mathematical and General*, **24**(7), L363.

Manna, S., Kiss, L. B., and Kertész, J. (1990). Cascades and self-organized criticality. *Journal of statistical physics*, **61**(3), 923–932.

Marcos, J., Vives, E., Mañosa, L., Acet, M., Duman, E., Morin, M., Novák, V., and Planes, A. (2003). Disorder-induced critical phenomena in magnetically glassy cu-al-mn alloys. *Phys. Rev. B*, **67**(22), 224406.

Marliére, C., Prades, S., Célarié, F., Dalmas, D., Bonamy, D., Guillot, C., and Bouchaud, E. (2003). overview on glass rough fracture. *J. Phys.: Condensed Matter*, **15**, S2377.

Marone, C. (1998). Laboratory-derived friction laws and their application to seismic faulting. *Annu. Rev. Earth Planet*, **26**, 643.

Martin, P. C., Siggia, E., and Rose, H. (1973). Statistical dynamics of classical systems. *Phys. Rev. A*, **8**(1), 423.

Martys, N., Cieplak, M., and Robbins, M. O. (1991*a*). Critical phenomena in fluid invasion of porous media. *Phys. Rev. Lett.*, **66**, 1058.

Martys, N., Cieplak, M., and Robbins, M. O. (1991*b*). Scaling relations for interface motion through disordered media: Application to two-dimensional fluid invasion. *Phys. Rev. B*, **1991**, 12294.

Maslov, S. (1995). Time directed avalanches in invasion models. *Phys. Rev. Lett.*, **74**(4), 562.

McMichael, R. D., Swartzendruber, L. J., and Bennet, L. H. (1993). Langevin approach to hysteresis and Barkhausen noise jump modelling in steel. *J. Appl. Phys.*, **73**, 5848–5850.

Mehta, A., Mills, A., Dahmen, K., and Sethna, J. (2002). Universal pulse shape scaling function and exponents: A critical test for avalanche models applied to Barkhausen noise. *Phys. Rev. E*, **65**, 046139.

Meisel, L. V. and Cote, P. J. (1992). Power laws, flicker noise, and the Barkhausen effect. *Phys. Rev. B*, **46**, 10822–10828.

Miguel, M., Andrade Jr, J. S., and Zapperi, S. (2003). Deblocking of interacting particle assemblies: From pinning to jamming. *Braz. J. Phys.*, **33**(3), 557–572.

Miguel, M.-C., Vespignani, A., Zaiser, M., and Zapperi, S. (2002). Dislocation Jamming and Andrade Creep. *Phys. Rev. Lett.*, **89**(16), 165501.

Miguel, M.-C., Vespignani, A., Zapperi, S., Weiss, J., and Grasso, J.-R. (2001). Intermittent dislocation flow in viscoplastic deformation. *Nature*, **410**, 667–671.

Miller, B., O'Hern, C., and Behringer, R. P. (1996). Stress fluctuations for continously sheared granular materials. *Phys. Rev. Lett.*, **77**, 3110–3113.

Minozzi, M., Caldarelli, G., Pietronero, L., and Zapperi, S. (2003). Dynamic fracture model for acoustic emission. *Eur. Phys. J. B*, **36**, 203.

Monette, L. and Anderson, M. P. (1994). Elastic and fracture properties of the two-dimensional triangular and square lattices. *Modelling Simul. Mater. Sci. Eng.*, **2**, 53.

Monier, D. and Fruchter, L. (2000). Fluid and plastic flow dynamics of the critical state for a strongly pinned 2d superconductor. *Eur. Phys. J. B*, **17**, 201.

Moreira, A. A., Andrade, J. S., Filho, J. M., and Zapperi, S. (2002). Boundary effects on flux penetration in disordered superconductors. *Phys. Rev. B*, **66**, 174507.

Moretti, P., Miguel, M.-C., Zaiser, M., and Zapperi, S. (2004). Depinning transition of dislocation assemblies: Pileups and low-angle grain boundaries. *Phys. Rev. B*, **69**(21), 214103.

Müller, M. and Wyart, M. (2015). Marginal stability in structural, spin, and electron glasses. *Annu. Rev. Condens. Matter Phys.*, **6**(1), 177–200.

Munoz, M. A. (2018). Colloquium: Criticality and dynamical scaling in living systems. *Rev. Mod. Phys.*, **90**(3), 031001.

Nagel, S. R. (1992). Instabilities in a sandpile. *Rev. Mod. Phys.*, **64**(1), 321–325.

Narayan, O. (1996). Self-similar Barkhausen noise in magnetic domain wall motion. *Phys. Rev. Lett.*, **77**, 3855–3858.

Narayan, O. and Fisher, D. S. (1993). Threshold critical dynamics of driven interfaces in random media. *Phys. Rev. B*, **48**, 7030–7042.

Nasuno, S., Kudrolli, A., Bank, A., and Gollub, J. P. (1998). Time-resolved studies of stick-slip friction in sheared granular layers. *Phys. Rev. E*, **58**(2), 2161–2171.

Nasuno, S., Kudrolli, A., and Gollub, J. (1997). Friction in granular layers: Hysteresis and precursors. *Phys. Rev. Lett.*, **79**, 949.

Nattermann, T. (1983). Interface phenomenology, dipolar interaction, and the dimensionality dependence of the incommensurate-commensurate transition. *J. Phys. C*, **16**, 4125–4135.

Nattermann, T., Stepanow, S., Tang, L. H., and Leschhorn, H. (1992). Dynamics of interface depinning in a disordered medium. *J. Phys. II (France)*, **2**, 1483–1488.

Néel, L. (1946). Bases d'une nouvelle theorie generale du champ coercitif. *Ann. Univ. Grenoble*, **22**, 299–343.

Néel, L. (1955). Energie des parois de Bloch dans le couches minces. *C. R. Acad. Sci. Paris*, **241**, 533–536.

Newman, M. (1996). Self-organized criticality, evolution and the fossil extinction record. *Proc. Royal Soc. of London. Series B: Biological Sciences*, **263**(1376), 1605–1610.

Nicolas, A., Ferrero, E. E., Martens, K., and Barrat, J.-L. (2018). Deformation and flow of amorphous solids: Insights from elastoplastic models. *Rev. Mod. Phys.*, **90**(4), 045006.

Nooruddin, H. A. and Blunt, M. J. (2018). Large-scale invasion percolation with trapping for upscaling capillary-controlled Darcy-scale flow. *Transport in Porous Media*, **121**(2), 479–506.

Novoselov, K. S., Geim, A. K., Dubonos, S. V., Hill, E. W., and Grigoreva, I. V. (2003). Subatomic movements of a domain wall in the Peierls potential. *Nature*, **426**, 812–816.

Nukala, P. K. V. V., Zapperi, S., and Simunovic, S. (2005). Statistical properties of fracture in a random spring model. *Phys. Rev. E*, **71**, 066106.

Ódor, G. (2004). Universality classes in nonequilibrium lattice systems. *Rev. Mod. Phys.*, **76**(3), 663.

Olson, C. J., Reichhardt, C., Groth, J., Field, S. B., and Nori, F. (1997*a*). Plastic flow, voltage noise and vortex avalanches in superconductors. *Physica C*, **290**, 89.

Olson, C. J., Reichhardt, C., and Nori, F. (1997*b*). Superconducting vortex avalanches, voltage bursts, and vortex plastic flow: Effect of the microscopic pinning landscape on the macroscopic properties. *Phys. Rev. B*, **56**, 6175.

Ovaska, M., Lehtinen, A., Alava, M. J., Laurson, L., and Zapperi, S. (2017). Excitation spectra in crystal plasticity. *Phys. Rev. Lett.*, **119**(26), 265501.

Paczuski, M. and Boettcher, S. (1996). Universality in sandpiles, interface depinning, and earthquake models. *Phys. Rev. Lett.*, **77**(1), 111–114.

Paczuski, M., Maslov, S., and Bak, P. (1996). Avalanche dynamics in evolution, growth, and depinning models. *Phys. Rev. E*, **53**(1), 414–443.

Paniz Mondolfi, A. E., Jour, G., Johnson, M., Reidy, J., Cason, R. C., Barkoh, B. A., Benaim, G., Singh, R., and Luthra, R. (2013). Primary cutaneous carcinosarcoma: Insights into its clonal origin and mutational pattern expression analysis through next-generation sequencing. *Hum Pathol*, **44**(12), 2853–60.

Papadopoulos, L., Puckett, J. G., Daniels, K. E., and Bassett, D. S. (2016). Evolution of network architecture in a granular material under compression. *Phys. Rev. E*, **94**(3), 032908.

Papanikolaou, S., Bohn, F., Sommer, R. L., Durin, G., Zapperi, S., and Sethna, J. P. (2011). Universality beyond power laws and the average avalanche shape. *Nat. Phys.*, **7**(4), 316–320.

Papanikolaou, S., Dimiduk, D. M., Choi, W., Sethna, J. P., Uchic, M. D., Woodward, C. F., and Zapperi, S. (2012). Quasi-periodic events in crystal plasticity and the self-organized avalanche oscillator. *Nature*, **490**(7421), 517–521.

Park, J.-A., Kim, J. H., Bi, D., Mitchel, J. A., Qazvini, N. T., Tantisira, K., Park, C. Y., McGill, M., Kim, S.-H., Gweon, B., Notbohm, J., Steward, Jr, R., Burger, S., Randell, S. H., Kho, A. T., Tambe, D. T., Hardin, C., Shore, S. A., Israel, E., Weitz, D. A., Tschumperlin, D. J., Henske, E. P., Weiss, S. T., Manning, M. L., Butler, J. P., Drazen, J. M., and Fredberg, J. J. (2015). Unjamming and cell shape in the asthmatic airway epithelium. *Nat. Mater.*, **14**, 1040–1048.

Pázmándi, F., Zaránd, G., and Zimányi, G. T. (1999). Self-organized criticality in the hysteresis of the Sherrington-Kirkpatrick model. *Phys. Rev. Lett.*, **83**(5), 1034.

Perkovic, O., Dahmen, K., and Sethna, J. P. (1995). Avalanches, Barkhausen noise, and plain old criticality. *Phys. Rev. Lett.*, **75**, 4528–4531.

Perkovic, O., Dahmen, K. A., and Sethna, J. P. (1999). Disorder-induced critical phenomena in hysteresis: Numerical scaling in three and higher dimensions. *Phys. Rev. B*, **59**, 6106–6119.

Petermann, T., Thiagarajan, T. C., Lebedev, M. A., Nicolelis, M. A., Chialvo, D. R., and Plenz, D. (2009). Spontaneous cortical activity in awake monkeys composed of neuronal avalanches. *PNAS*, **106**(37), 15921–15926.

Petri, A., Paparo, G., Vespignani, A., Alippi, A., and Costantini, M. (1994). Experimental evidence for critical dynamics in microfracturing processes. *Phys. Rev. Lett.*, **73**, 3423.

Phoenix, S. L., Ibnabdeljalil, M., and Hui, C.-Y. (1997). Size effects in the distribution for strength of brittle matrix fibrous composites. *Int. J. Solids Structures*, **34**, 545.

Phoenix, S. L. and Ra, R. (1992). Scalings in fracture probabilities for a brittle matrix fiber composite. *Acta Metall. Mater.*, **40**, 2813.

Pierce, F. T. (1926). Tensile tests for cotton yarns. *J. Textile Inst.*, **17**, 355.

Pietronero, L., Vespignani, A., and Zapperi, S. (1994). Renormalization scheme for self-organized criticality in sandpile models. *Phys. Rev. Lett.*, **72**(11), 1690.

Pla, O. and Nori, F. (1991). Self-organized critical behavior in pinned flux lattices. *Phys. Rev. Lett.*, **67**, 919.

Plenz, D., Ribeiro, T. L., Miller, S. R., Kells, P. A., Vakili, A., and Capek, E. L. (2021). Self-organized criticality in the brain. *Frontiers in Physics*, **9**, 365.

Politi, A., Ciliberto, S., and Scorretti, R. (2002). Failure time in the fiber-bundle model with thermal noise and disorder. *Phys. Rev. E*, **66**, 026107.

Ponson, L., Bonamy, D., and Bouchaud, E. (2006). Two-dimensional scaling properties of experimental fracture surfaces. *Phys. Rev. Lett.*, **96**, 035506.

Poujade, M., Grasland-Mongrain, E., Hertzog, A., Jouanneau, J., Chavrier, P., Ladoux, B., Buguin, A., and Silberzan, P. (2007). Collective migration of an epithelial monolayer in response to a model wound. *PNAS*, **104**(41), 15988–93.

Prades, S., Bonamy, D., Dalmas, D., Bouchaud, E., and Guillot, C. (2004). Nanoductile crack propagation in glasses under stress corrosion: spatiotemporal evolution of damage in the vicinity of the crack tip. *Int. J. Sol. Struct.*, **42**, 637.

Priezzhev, V. B. (1994). Structure of two-dimensional sandpile. i. height probabilities. *Journal of statistical physics*, **74**(5), 955–979.

Puppin, E. (2000). Statistical properties of Barkhausen noise in thin fe films. *Phys. Rev. Lett.*, **84**, 5415–5418.

Puppin, E., Vavassori, P., and Callegaro, L. (2000). A focused magneto-optical Kerr magnetometer for Barkhausen jump observations. *Rev. Sci. Instr.*, **71**, 1752–1755.

Räisänen, V. I., Alava, M. J., and Nieminen, R. M. (1998). Fracture of three-dimensional fuse networks with quenched disorder. *Phys. Rev. B*, **58**, 14288.

Ramamurty, U., Jana, S., Kawamura, Y., and Chattopadhyay, K. (2005). Hardness and plastic deformation in a bulk metallic glass. *Acta Materialia*, **53**(3), 705 – 717.

Ramanathan, S., Ertas, D., and Fisher, D. S. (1997). Quasistatic crack propagation in heterogeneous media. *Phys. Rev. Lett.*, **79**, 873.

Ramanathan, S. and Fisher, D. S. (1997). Dynamics and instabilities of planar tensile cracks in heterogeneous media. *Phys. Rev. Lett.*, **79**, 877.

Ramanathan, S. and Fisher, D. S. (1998). Onset of propagation of planar cracks in heterogeneous media. *Phys. Rev. B*, **58**, 6026.

Ramaswamy, S. (2010). The mechanics and statistics of active matter. *Annual Review of Condensed Matter Physics*, **1**(1), 323–345.

Reichhardt, C., Olson, C., Groth, J., Field, S., and Nori, F. (1996). Vortex plastic flow, local flux density, magnetization hysteresis loops, and critical current, deep in the Bose-glass and Mott-insulator regimes. *Phys. Rev. B*, **53**(14), R8898.

Reynolds, O. (1885). On the dilatancy of media composed of rigid particles in contact.

Phil. Mag., **20**, 469.

Richeton, T., Weiss, J., and Louchet, F. (2005). Breakdown of avalanche critical behaviour in polycrystalline plasticity. *Nature Materials*, **4**, 465–469.

Rolley, E., Guthmann, C., Gombrowicz, R., and Repain, V. (1998). Roughness of the contact line on a disordered substrate. *Phys. Rev. Lett.*, **80**(13), 2865–2868.

Rørth, P. (2009). Collective cell migration. *Annu. Rev. Cell Dev. Biol.*, **25**, 407–29.

Rosendahl, J., Vekić, M., and Kelley, J. (1993). Persitent self-organization of sandpiles. *Phys. Rev. E*, **47**(2), 1401.

Rosendahl, J., Vekic, M., and Rutledge, J. (1994). Predictability of large avalanches on a sandpile. *Phys. Rev. Lett.*, **73**(4), 537–540.

Rosinberg, M. L., Kierlik, E., and Tarjus, G. (2003). Percolation depinning and and avalanches in capillary condensation of gases in disordered porous solids. *Europhys. Lett.*, **62**, 377.

Rossi, M., Pastor-Satorras, R., and Vespignani, A. (2000). Universality class of absorbing phase transitions with a conserved field. *Phys. Rev. Lett.*, **85**(9), 1803.

Rosso, A., Hartmann, A. K., and Krauth, W. (2003). Depinning of elastic manifolds. *Phys. Rev. E*, **67**, 021602.

Rosso, A. and Krauth, W. (2001). Origin of the roughness exponent in elastic strings at the depinning threshold. *Phys. Rev. Lett.*, **87**(18), 187002.

Rosso, A. and Krauth, W. (2002). Roughness at the depinning threshold for a long-range elastic string. *Phys. Rev. E*, **65**, 025101.

Rosso, A., Le Doussal, P., and Wiese, K. J. (2009). Avalanche-size distribution at the depinning transition: A numerical test of the theory. *Phys. Rev. B*, **80**, 144204.

Roux, S. (2000). Thermally activated breakdown in the fiber-bundle model. *Phys. Rev. E*, **62**, 6164.

Roux, S. and Guyon, E. (1985). Mechanical percolation: A small beam lattice study. *J. Physique Lett.*, **46**, L999.

Roux, S. and Guyon, E. (1989). Temporal development of invasion percolation. *J. Phys. A: Mathematical and General*, **22**(17), 3693–3705.

Roux, S., Hansen, A., Herrmann, H., and Guyon, E. (1988). Rupture of heterogeneous media in the limit of infinite disorder. *Journal of statistical physics*, **52**(1), 237–244.

Ruelle, P. and Sen, S. (1992). Toppling distributions in one-dimensional Abelian sandpiles. *J. Phys. A: Mathematical and General*, **25**(22), L1257.

Rundle, J. B. and Klein, W. (1989). Nonclassical nucleation and growth of cohesive tensile cracks. *Phys. Rev. Lett.*, **63**, 171.

Ryu, K.-S., Akinaga, H., and Shin, S.-C. (2007). Tunable scaling behaviour observed in Barkhausen criticality of a ferromagnetic film. *Nat. Phys.*, **3**(8), 547–550.

Sahimi, M. and Arbabi, S. (1993). Mechanics of disordered solids. ii. percolation on elastic networks with bond bending forces. *Phys. Rev. B*, **47(2)**, 703–712.

Salerno, K. M., Maloney, C. E., and Robbins, M. O. (2012). Avalanches in strained amorphous solids: does inertia destroy critical behavior? *Phys. Rev. Lett.*, **109**(10), 105703.

Salerno, K. M. and Robbins, M. O. (2013). Effect of inertia on sheared disordered solids: Critical scaling of avalanches in two and three dimensions. *Phys. Rev. E*, **88**(6), 062206.

Salminen, L., Tolvanen, A., and Alava, M. J. (2002). Acoustic emission from paper fracture. *Physical Review Letters*, **89**(18), 185503.

Santucci, S., Grob, M., R.Toussaint, Schmittbuhl, J., Hansen, A., and Maloy, K. J. (2009). Crackling dynamics during the failure of heterogeneous material: Optical and acoustic tracking of slow interfacial crack growth. In *Proceedings of the 12th International Conference on Fracture, Ottawa*.

Schmittbuhl, J., Hansen, A., and Batrouni, G. G. (2003). Roughness of interfacial crack fronts: Stress-weighted percolation in the damage zone. *Phys. Rev. Lett.*, **90**, 045505.

Schmittbuhl, J. and Maloy, K. J. (1997). Direct observation of a self-affine crack propagation. *Phys. Rev. Lett.*, **78**, 3888.

Schmittbuhl, J., Roux, S., Villotte, J. P., and Maloy, K. J. (1995). Interfacial crack pinning: Effect of nonlocal interactions. *Phys. Rev. Lett.*, **74**, 1787.

Schöllmann, S. (1999). Simulation of a two-dimensional shear cell. *Phys. Rev. E*, **59**(1), 889–899.

Schuh, C. A., Hufnagel, T. C., and Ramamurty, U. (2007). Mechanical behavior of amorphous alloys. *Acta Materialia*, **55**(12), 4067–4109.

Schwarz, A., Liebmann, M., Kaiser, U., Wiesendanger, R., Noh, T. W., and Kim, D. W. (2004). Visualization of the Barkhausen effect by magnetic force microscopy. *Phys. Rev. Lett.*, **92**, 077206.

Schwerdtfeger, J., Nadgorny, E., Madani-Grasset, F., Koutsos, V., Blackford, J. R., and Zaiser, M. (2007). Scale-free statistics of plasticity-induced surface steps on KCl single crystals. *JSTAT*, **4**, 1.

Scialdone, A., Tanaka, Y., Jawaid, W., Moignard, V., Wilson, N. K., Macaulay, I. C., Marioni, J. C., and Göttgens, B. (2016). Resolving early mesoderm diversification through single-cell expression profiling. *Nature*, **535**(7611), 289–293.

Scorretti, R., Ciliberto, S., and Guarino, A. (2001). Disorder enhances the effects of thermal noise in the fiber bundle model. *Europhys. Lett.*, **55**, 626.

Selinger, R. L. B., Wang, Z.-G., and Gelbart, W. M. (1991a). Effect of temperature and small-scale defects on the strength of solids. *J. Chem. Phys*, **95**, 9128.

Selinger, R. L. B., Wang, Z.-G., Gelbart, W. M., and Ben-Saul, A. (1991b). Statistical-thermodynamic approach to fracture. *Phys. Rev. A*, **43**, 4396.

Sepúlveda, N., Petitjean, L., Cochet, O., Grasland-Mongrain, E., Silberzan, P., and Hakim, V. (2013). Collective cell motion in an epithelial sheet can be quantitatively described by a stochastic interacting particle model. *PLoS Comput Biol*, **9**(3), e1002944.

Serra, R., Villani, M., Graudenzi, A., and Kauffman, S. (2007). Why a simple model of genetic regulatory networks describes the distribution of avalanches in gene expression data. *Journal of theoretical biology*, **246**(3), 449–460.

Serra, R., Villani, M., and Semeria, A. (2004). Genetic network models and statistical properties of gene expression data in knock-out experiments. *Journal of theoretical biology*, **227**(1), 149–157.

Serra-Picamal, X., Conte, V., Vincent, R., Anon, E., Tambe, D. T., Bazellieres, E., Butler, J. P., Fredberg, J. J., and Trepat, X. (2012). Mechanical waves during tissue expansion. *Nat. Phys.*, **8**(8), 628–634.

Sethna, J., Dahmen, K. A., and Myers, C. R. (2001). Crackling noise. *Nature*, **410**, 242–244.

Sethna, J. P., Bierbaum, M. K., Dahmen, K. A., Goodrich, C. P., Greer, J. R., Hayden, L. X., Kent-Dobias, J. P., Lee, E. D., Liarte, D. B., Ni, X. et al. (2017). Deformation of crystals: Connections with statistical physics. *Annual Review of Materials Research*, **47**, 217–246.

Sethna, J. P., Dahmen, K., Kartha, S., Krumhansl, J. A., Roberts, B. W., and Shore, J. D. (1993). Hysteresis and hierarchies: Dynamics of disorder-driven first-order phase transformations. *Phys. Rev. Lett.*, **70**, 3347–3350.

Shan, Z., Li, J., Cheng, Y., Minor, A., Asif, S. S., Warren, O., and Ma, E. (2008). Plastic flow and failure resistance of metallic glass: Insight from in situ compression of nanopillars. *Phys. Rev. B*, **77**(15), 155419.

Sharoni, A., Ramírez, J. G., and Schuller, I. K. (2008). Multiple avalanches across the metal-insulator transition of vanadium oxide nanoscaled junctions. *Phys. Rev. Lett.*, **101**(2), 026404.

Shekhawat, A., Papanikolaou, S., Zapperi, S., and Sethna, J. P. (2011). Dielectric breakdown and avalanches at nonequilibrium metal-insulator transitions. *Phys. Rev. Lett.*, **107**(27), 276401.

Shekhawat, A., Zapperi, S., and Sethna, J. P. (2013). From damage percolation to crack nucleation through finite size criticality. *Phys. Rev. Lett.*, **110**(18), 185505.

Sheppard, A. P., Knackstedt, M. A., Pinczewski, W. V., and Sahimi, M. (1999). Invasion percolation: New algorithms and universality classes. *J. Phys. A: Mathematical and General*, **32**(49), L521.

Sherrington, D. and Kirkpatrick, S. (1975). Solvable model of a spin-glass. *Phys. Rev. Lett.*, **35**(26), 1792.

Shi, Y. and Falk, M. L. (2007). Stress-induced structural transformation and shear banding during simulated nanoindentation of a metallic glass. *Acta Materialia*, **55**(13), 4317 – 4324.

Shukla, P. (2000). Exact solution of return hysteresis loops in a one-dimensional random-field ising model at zero temperature. *Phys. Rev. E*, **62**, 4725.

Shukla, P. (2001). Exact expressions for minor hysteresis loops in the random field Ising model on a Bethe lattice at zero temperature. *Phys. Rev. E*, **63**, 027102.

Shumway, J. and Satpathy, S. (1997). Dynamics of flux penetration and critical currents in type-II superconductors. *Phys. Rev. B*, **56**, 103.

Smith, R. L. (1980). A probability model for fibrous composites with local load sharing. *Proc. R. Soc. London A*, **372**, 179.

Smith, R. L. and Phoenix, S. L. (1981). Asymptotic distributions for the failure of fibrous materials under series-parallel structure and equal load-sharing. *J. Appl. Mech.*, **48**, 75.

Sneppen, K. (1992). Self-organized pinning and interface growth in a random medium. *Phys. Rev. Lett.*, **69**(24), 3539–3542.

Sneppen, K., Bak, P., Flyvbjerg, H., and Jensen, M. H. (1995). Evolution as a self-organized critical phenomenon. *PNAS*, **92**(11), 5209–5213.

Sole, R. V. and Bascompte, J. (1996). Are critical phenomena relevant to large-scale evolution? *Proc. Royal Soc. of London. Series B: Biological Sciences*, **263**(1367),

161–168.

Solé, R. V. and Manrubia, S. C. (1996). Extinction and self-organized criticality in a model of large-scale evolution. *Phys. Rev. E*, **54**(1), R42.

Sole, R. V., Manrubia, S. C., Benton, M., and Bak, P. (1997). Self-similarity of extinction statistics in the fossil record. *Nature*, **388**(6644), 764–767.

Sornette, D. (1992*a*). Critical phase transitions made self-organized: a dynamical system feedback mechanism for self-organized criticality. *J. Phys. I (France)*, **2**(11), 2065–2073.

Sornette, D. (1992*b*). Mean-field solution of a block spring model of earthquakes. *J. Phys. I France*, **2**, 2089.

Sornette, D. (1994). Sweeping of an instability: An alternative to self-organized criticality to get power laws without parameter tuning. *J.Phys. I (France)*, **4**(2), 209–221.

Spasojevic, D., Bukvic, S., Milosevic, S., and Stanley, H. E. (1996). Barkhausen noise: elementary signals, power laws, and scaling relations. *Phys. Rev. E*, **54**, 2531–2546.

Stauffer, D. and Aharony, A. (1994). *Introduction to Percolation Theory*. Revised second edition, Taylor and Francis.

Steinway, S. N., Zañudo, J. G., Ding, W., Rountree, C. B., Feith, D. J., Loughran, T. P., and Albert, R. (2014). Network modeling of TGF-beta signaling in hepatocellular carcinoma epithelial-to-mesenchymal transition reveals joint Sonic Hedgehog and Wnt pathway activation. *Cancer Research*, **74**(21), 5963–5977.

Steinway, S. N., Zañudo, J. G. T., Michel, P. J., Feith, D. J., Loughran, T. P., and Albert, R. (2015). Combinatorial interventions inhibit TGFβ-driven epithelial-to-mesenchymal transition and support hybrid cellular phenotypes. *NPJ Syst. Biol. Appl.*, **1**, 15014.

Su, C. and Anand, L. (2006). Plane strain indentation of a Zr-based metallic glass: experiments and numerical simulation. *Acta Materialia*, **54**, 179–189.

Sujeer, M. K., Buldyrev, S. V., Zapperi, S., Jose S. Andrade, J., Stanley, H. E., and Suki, B. (1997). Volume distributions of avalanches in lung inflation: A statistical mechanical approach. *Phys. Rev. E*, **56**(3), 3385–3394.

Suki, B., Alencar, A. M., Tolnai, J., Asztalos, T., Petak, F., Sujeer, M. K., Patel, K., Patel, J., Stanley, H. E., and Hantos, Z. (2000). Size distribution of recruited alveolar volumes in airway reopening. *J Appl Physiol*, **89**, 2030–2040.

Suki, B., Andrade, J. S., Coughlin, M., Stamenovic, D., Stanley, H., Sujeer, M., and Zapperi, S. (1998). Mathematical modeling of the first inflation of degassed lungs. *Ann. Biomed. Eng.*, **26**, 608.

Suki, B., Barabási, A.-L., Hantos, Z., Peták, F., and Stanley, H. E. (1994). Avalanches and power-law behaviour in lung inflation. *Nature*, **368**, 615–8.

Sun, B., Yu, H., Jiao, W., Bai, H., Zhao, D., and Wang, W. (2010). Plasticity of ductile metallic glasses: A self-organized critical state. *Phys. Rev. Lett.*, **105**(3), 035501.

Surdeanu, R., Wijngaarden, R., Visser, E., Huijbregtse, J., Rector, J., Dam, B., and Griessen, R. (1999). Kinetic roughening of penetrating flux fronts in high-Tc thin film superconductors. *Phys. Rev. Lett.*, **83**(10), 2054.

Szabó, B., Szöllösi, G. J., Gönci, B., Jurányi, Z., Selmeczi, D., and Vicsek, T. (2006).

Phase transition in the collective migration of tissue cells: Experiment and model. *Phys. Rev. E*, **74**, 061908.

Tadić, B., Nowak, U., Usadel, K.-D., Ramaswamy, R., and Padlewski, S. (1992). Scaling behavior in disordered sandpile automata. *Phys. Rev. A*, **45**(12), 8536.

Talamali, M., Petäjä, V., Vandembroucq, D., and Roux, S. (2011). Avalanches, precursors, and finite-size fluctuations in a mesoscopic model of amorphous plasticity. *Phys. Rev. E*, **84**(1), 016115.

Tallakstad, K. T., Toussaint, R., Santucci, S., Schmittbuhl, J., and Maloy, K. J. (2011). Local dynamics of a randomly pinned crack front during creep and forced propagation: an experimental study. *Phys Rev E Stat Nonlin Soft Matter Phys*, **83**(4 Pt 2), 046108.

Tambe, D. T., Hardin, C. C., Angelini, T. E., Rajendran, K., Park, C. Y., Serra-Picamal, X., Zhou, E. H., Zaman, M. H., Butler, J. P., Weitz, D. A., Fredberg, J. J., and Trepat, X. (2011). Collective cell guidance by cooperative intercellular forces. *Nat. Mater.*, **10**(6), 469–75.

Tang, C. and Bak, P. (1988*a*). Critical exponents and scaling relations for self-organized critical phenomena. *Phys. Rev. Lett.*, **60**(23), 2347.

Tang, C. and Bak, P. (1988*b*). Mean field theory of self-organized critical phenomena. *Journal of Statistical Physics*, **51**(5), 797–802.

Tang, L.-H. and Leschhorn, H. (1992). Pinning by directed percolation. *Phys. Rev. A*, **45**(12), R8309–R8312.

Tanguy, A., Gounelle, M., and Roux, S. (1998). From individual to collective pinning: Effect of long-range elastic interactions. *Phys. Rev. E*, **58**, 1577–1590.

Taylor, D. W. (1945). *Fundamentals of soil mechanics*. New York, Wiley.

Thompson, A. H., Katz, A. J., and Raschke, R. A. (1987). Mercury injection in porous media: A resistance devil's staircase with percolation geometry. *Phys. Rev. Lett.*, **58**(1), 29–32.

Thompson, P. A. and Grest, G. S. (1991). Granular flow: Friction and the dilatancy transition. *Phys. Rev. Lett.*, **67**(13), 1751–1754.

Toussaint, R. and Pride, S. R. (2005). Interacting damage models mapped onto Ising and percolation models. *Phys. Rev. E*, **71**, 046127.

Travers, T., Ammi, M., Bideau, D., Gervois, A., Messager, J.-C., and Troadec, J.-P. (1987). Uniaxial compression of 2d packings of cylinders. Effects of weak disorder. *Europhys. Lett.*, **4**(3), 329–332.

Trexler, M. M. and Thadhani, N. N. (2010). Mechanical properties of bulk metallic glasses. *Progress in Materials Science*, **55**(8), 759–839.

Tripathi, S., Kessler, D. A., and Levine, H. (2020). Biological networks regulating cell fate choice are minimally frustrated. *Phys. Rev. Lett.*, **125**(8), 088101.

Uchic, M. D., Dimiduk, D. M., Florando, J. N., and Nix, W. D. (2004). Sample Dimensions Influence Strength and Crystal Plasticity. *Science*, **305**, 986–989.

Urbach, J. S., Madison, R. C., and Markert, J. T. (1995*a*). Interface depinning, self-organized criticality, and the Barkhausen effect. *Phys. Rev. Lett.*, **75**, 276–279.

Urbach, J. S., Madison, R. C., and Markert, J. T. (1995*b*). Reproducible noise in a macroscopic system: Magnetic avalanches in Permivar. *Phys. Rev. Lett.*, **75**, 4694–4697.

Vaidyanathan, R., Dao, M., Ravichandran, G., and Suresh, S. (2001). Study of mechanical deformation in bulk metallic glass through instrumented indentation. *Acta Materialia*, **49**(18), 3781 – 3789.

Vedula, S. R. K., Ravasio, A., Lim, C. T., and Ladoux, B. (2013). Collective cell migration: A mechanistic perspective. *Physiology (Bethesda)*, **28**(6), 370–9.

Veje, C. T., Howell, D. W., and Behringer, R. P. (1999). Kinematics of a 2D granular Couette experiment at the transition to shearing. *Phys. Rev. E*, **59**(1), 739–745.

Vergeles, M., Maritan, A., and Banavar, J. R. (1997). Mean-field theory of sandpiles. *Phys. Rev. E*, **55**(2), 1998.

Vespignani, A., Dickman, R., Munoz, M. A., and Zapperi, S. (1998). Driving, conservation, and absorbing states in sandpiles. *Phys. Rev. Lett.*, **81**(25), 5676.

Vespignani, A., Dickman, R., Munoz, M. A., and Zapperi, S. (2000). Absorbing-state phase transitions in fixed-energy sandpiles. *Phys. Rev. E*, **62**(4), 4564.

Vespignani, A. and Zapperi, S. (1997). Order parameter and scaling fields in self-organized criticality. *Phys. Rev. Lett.*, **78**(25), 4793.

Vespignani, A. and Zapperi, S. (1998). How self-organized criticality works: A unified mean-field picture. *Phys. Rev. E*, **57**(6), 6345.

Vespignani, A., Zapperi, S., and Pietronero, L. (1995). Renormalization approach to the self-organized critical behavior of sandpile models. *Phys. Rev. E*, **51**(3), 1711.

Vives, E. and Planes, A. (1994). Avalanches in a fluctuationless phase transition in a random-bond Ising model. *Phys. Rev. B*, **50**, 3839–3848.

Vives, E. and Planes, A. (2000). Hysteresis and avalanches in disordered systems. *J. Mag. Magn. Mat.*, **221**, 164–171.

Vives, E. and Planes, A. (2001). Hysteresis and avalanches in the random anisotropy Ising model. *Phys. Rev. B*, **63**, 134431.

Vlasko-Vlasov, V. K., Welp, U., Metlushko, V., and Crabtree, G. W. (2004). Experimental test of the self-organized criticality of vortices in superconductors. *Phys. Rev. B*, **69**(14), 140504.

Waddington, C. (1957). *The Strategy of the Genes*. Allen & Unwin, London.

Wang, G., Chan, K. C., Xia, L., Yu, P., Shen, J., and Wang, W. (2009). Self-organized intermittent plastic flow in bulk metallic glasses. *Acta Materialia*, **57**(20), 6146–6155.

Wang, J., Xu, L., Wang, E., and Huang, S. (2010). The potential landscape of genetic circuits imposes the arrow of time in stem cell differentiation. *Biophys. J.*, **99**(1), 29–39.

Wang, J., Zhang, K., Xu, L., and Wang, E. (2011). Quantifying the Waddington landscape and biological paths for development and differentiation. *Proc Natl Acad Sci USA*, **108**(20), 8257–62.

Wang, Z.-G., Landman, U., Selinger, R. L. B., and Gelbart, W. M. (1991). Molecular-dynamics study of elasticity and failure of ideal solids. *Phys. Rev. B*, **44**, 378.

Weeks, E. R., Crocker, J. C., Levitt, A. C., Schofield, A., and Weitz, D. A. (2000). Three-dimensional direct imaging of structural relaxation near the colloidal glass transition. *Science*, **287**, 627–631.

Weiss, J. (1997). The role of attenuation on acoustic emission amplitude distributions and b-values. *Bull. Seism. Soc. Am.*, **87**, 1362.

Weiss, J., Lahaie, F., and Grasso, J. R. (2000). Statistical analysis of dislocation dynamics during viscoplastic deformation from acoustic emission. *J. Geophys. Res.*, **105**, 433–442.

Weiss, J. and Marsan, D. (2003). Three-dimensional mapping of dislocation avalanches: Clustering and space/time coupling. *Science*, **299**, 89–92.

Welling, M., Aegerter, C., and Wijngaarden, R. (2004). Noise correction for roughening analysis of magnetic flux profiles in $yba_2cu_3o_{7-x}$. *Eur. Phys. J. B*, **38**(1), 93–98.

Welling, M. S., Aegerter, C. M., and Wijngaarden, R. J. (2005). Self-organized criticality induced by quenched disorder: Experiments on flux avalanches in NbH_x films. *Phys. Rev. B*, **71**(10), 104515.

White, R. A. and Dahmen, K. A. (2003). Driving rate effects on crackling noise. *Phys. Rev. Lett.*, **91**(8), 085702.

Wiegman, N. J. (1977). Barkhausen effect in magnetic thin films: experimental noise spectra. *Appl. Phys.*, **12**, 157–161.

Wiegman, N. J. (1979). Barkhausen noise in magnetic thin films. Master's thesis, Technical University, Eindhoven. Available online: http://alexandria.tue.nl/extra3/proefschrift/PRF3A/7909139.pdf.

Wiegman, N. J. and Stege, R. (1978). Barkhausen effect in magnetic thin films: general behaviour and stationarity along the hysteresis loop. *Appl. Phys.*, **16**, 167–174.

Wilkens, M. (1976). Broadening of x-ray diffraction lines of crystals containing dislocation distributions. *Kristall und Technik*, **11**(11), 1159–1169.

Wilkinson, D. and Willemsen, J. F. (1983). Invasion percolation: a new form of percolation theory. *J. Phys. A: Mathematical and General*, **16**(14), 3365–3376.

Wood, D. (1990). *Soil Behaviour and Critical State Soil Mechanics*. Cambridge University Press, Cambridge.

Wootters, A. H. and Hallock, R. B. (2003). Superfluid avalanches in nuclepore: Constrained versus free-boundary experiments and simulations. *Phys. Rev. Lett.*, **91**(16), 165301.

Xu, L., Davies, S., Schofield, A. B., and Weitz, D. A. (2008). Dynamics of drying in 3d porous media. *Phys. Rev. Lett.*, **101**(9), 094502.

Yan, L., Baity-Jesi, M., Müller, M., and Wyart, M. (2015). Dynamics and correlations among soft excitations in marginally stable glasses. *Phys. Rev. Lett.*, **114**(24), 247208.

Zadeh, A. A., Barés, J., Brzinski, T. A., Daniels, K. E., Dijksman, J., Docquier, N., Everitt, H. O., Kollmer, J. E., Lantsoght, O., Wang, D. et al. (2019). Enlightening force chains: A review of photoelasticimetry in granular matter. *Granular Matter*, **21**(4), 1–12.

Zaiser, M. (2006). Scale invariance in plastic flow of crystalline solids. *Advances in Physics*, **55**, 185–245.

Zaiser, M., Grasset, F. M., Koutsos, V., and Aifantis, E. C. (2004). Self-Affine Surface Morphology of Plastically Deformed Metals. *Phys. Rev. Lett.*, **93**(19), 195507.

Zaiser, M. and Hähner, P. (1997). Oscillatory modes of plastic deformation: Theoretical concepts. *Phys. Stat. Sol. b*, **197**, 267.

Zaiser, M. and Moretti, P. (2005). Fluctuation phenomena in crystal plasticity: A continuum model. *JSTAT*, **8**, 4.

Zaitsev, S. I. (1992). Robin hood as self-organized criticality. *Physica A: Statistical and Theoretical Physics*, **189**(3-4), 411–416.

Zapperi, S., Castellano, C., Colaiori, F., and Durin, G. (2005a). Signature of effective mass in crackling-noise asymmetry. *Nature Phys.*, **1**, 46–49.

Zapperi, S., Cizeau, P., Durin, G., and Stanley, H. E. (1998). Dynamics of a ferromagnetic domain wall: avalanches, depinning transition and the Barkhausen effect. *Phys. Rev. B*, **58**, 6353–6366.

Zapperi, S., Herrmann, H. J., and Roux, S. (2000). Planar cracks in the fuse model. *Eur. Phys. J. B*, **17**, 131.

Zapperi, S., Lauritsen, K. B., and Stanley, H. E. (1995). Self-organized branching processes: Mean-field theory for avalanches. *Phys. Rev. Lett.*, **75**(22), 4071.

Zapperi, S., Moreira, A. A., and Andrade, J. S. (2001). Flux front penetration in disordered superconductors. *Phys. Rev. Lett.*, **86**(16), 3622–3625.

Zapperi, S., Nukala, P. K. V. V., and Simunovic, S. (2005b). Crack avalanches in the three-dimensional random fuse model. *Physica A*, **357**, 129.

Zapperi, S., Nukala, P. K. V. V., and Simunovic, S. (2005c). Crack roughness and avalanche precursors in the random fuse model. *Phys. Rev. E*, **71**, 026106.

Zapperi, S., Ray, P., Stanley, H. E., and Vespignani, A. (1997a). First-order transition in the breakdown of disordered media. *Phys. Rev. Lett.*, **78**, 1408.

Zapperi, S., Ray, P., Stanley, H. E., and Vespignani, A. (1999). Avalanches in breakdown and fracture processes. *Phys. Rev. E*, **59**, 5049.

Zapperi, S., Vespignani, A., and Stanley, H. E. (1997b). Plasticity and avalanche behaviour in microfracturing phenomena. *Nature*, **388**, 658.

Zeraati, R., Priesemann, V., and Levina, A. (2021). Self-organization toward criticality by synaptic plasticity. *Frontiers in Physics*, 103.

Zhang, S. and Ding, E. (1996). Failure of fiber bundles with local load sharing. *Phys. Rev. B*, **53**, 646.

Zhou, S. J. and Curtin, W. A. (1995). Failure of fiber composites: a lattice green function model. *Acta. Metal. Mater.*, **43**, 3093.

Index